The Best
AMERICAN
SCIENCE &
NATURE
WRITING
2024

T0282971

The Best
AMERICAN
SCIENCE &
NATURE
WRITING™
2024

Edited and with an Introduction
BY BILL MCKIBBEN

JAIME GREEN, *Series Editor*

MARINER BOOKS
New York   Boston

FIRST EDITION

ISSN 1530-1508

ISBN 978-0-06-333399-4

24 25 26 27 28 LBC 5 4 3 2 1

Contents

Foreword

EVERY FALL, I collect a list of recommendations for the incoming guest editor (this year, the wonderful Bill McKibben), a starting point for their readings and selections for the book. This year was the first time I found myself wondering whether all the publications whose work we're honoring would still exist by the time the book comes out.

It is essentially luck that every piece in this collection comes from a publication still extant at the time of my writing. Among the notable essays listed at the back of the book—almost a hundred excellent pieces—are pieces from publications that have essentially shuttered, magazines that laid off their science desks or their entire staff of writers. There are journalists who luckily avoided a wave of layoffs, and surely more who worry what the next year might bring. It's impossible to look back on the past year of science and nature writing without this sense of absence and eulogy.

This isn't the first time we've seen such cullings. *Pacific Standard*, which covered social and environmental justice with tenacity and elegance, stopped publication in 2019. *Pacific Standard* was even a nonprofit, a hopeful alternative to business models that leave magazines subject to market whims. But that meant it was subject to the whims of its funder. The winds changed, the funder withdrew, the magazine shut down. We have to consider ourselves lucky that the archives—which include several pieces featured in past issues of this collection—are still online at all, after *Grist* acquired the magazine.

We thought for a long time that the internet would kill journalism. In plenty of ways, it hasn't helped: Google's monopoly on search funnels and ads, Facebook's inflated stats that led the doomed pivot to video, the death of classifieds and other newsprint revenue sources. Readers often now expect to access journalism for free and balk at paywalls, seeking workarounds rather than paying for what they read, and websites are so clogged with ads as to be almost unreadable. Writers are still consistently paid far less for work published online than in print, even at the same publication, which seems to come down to little more than prestige and what publications can get away with.

We thought the internet would kill journalism. We didn't see the vultures coming.

They came for local newspapers first. The hedge fund Alden Global Capital innovated a model that McKay Coppins, writing in *The Atlantic* in 2021, called "strip-mining local-news outfits." He elaborated: "The model is simple: Gut the staff, sell the real estate, jack up subscription prices, and wring as much cash as possible out of the enterprise until eventually enough readers cancel their subscriptions that the paper folds, or is reduced to a desiccated husk of its former self." It didn't matter if a newspaper was struggling or stable when Alden bought it, the vicious playbook was the same.

As of 2019, Alden was running over a hundred newspapers, and they'd cut two of every three jobs. It's a model known as "vulture capitalism," though a former *Chicago Tribune* reporter who'd seen Alden's impact firsthand told Coppins the name didn't quite fit. "A vulture doesn't hold a wounded animal's head underwater. This is predatory."

The predators have come for online news, too. Gawker Media is a classic example, first undermined by Peter Thiel's vindictive, opportunistic funding of the lawsuit that would bankrupt the company. After a few years owned by Univision, the group of websites now called Gizmodo Media Group was sold to private equity firm Great Hill Partners. Megan Greenwell was editor-in-chief of Deadspin, GMG's sports and culture website. In 2019, upon her resignation, she wrote, "A metastasizing swath of media is controlled by private-equity vultures and capricious billionaires."

Newspapers and magazines have been reduced to figures on a spreadsheet. A business is a business—that part's not new—

but the men who run Alden Global Capital never even pretend to care about journalism. When they bought the *Chicago Tribune* in 2021, Coppins writes, "The new owners did not fly to Chicago to address the staff, nor did they bother with paeans to the vital civic role of journalism. Instead, they gutted the place." Greenwell wrote of GMG's owners: "They want a quick cash-out rather than the growth that comes from a well-run business." It's impossible to know how much science writing has been lost at the hundreds of local papers now owned by financial firms, by Alden and those aping their mercenary model.

But here's some of what we do know: 2023 was a bloodbath year for science journalism. *National Geographic*, owned by Disney, laid off its last staff writers; Gizmodo laid off its last climate writer (so we know what Great Hill Partners did there); CNBC disbanded its climate desk; and *Wired* laid off around twenty employees. According to SFGate, "The Wired layoffs hit reporters who covered labor, health, artificial intelligence, space and science and had written about such topics as AI regulation, autoworker organizing, new vaccine developments and more. Several editors also received pink slips, including those in the features and science divisions." Writers, editors, production staff—gone. And with perhaps the most dramatic sign of the state of the industry, *Popular Science*, which had just celebrated its 150th anniversary in 2022, ceased publication. *Popular Science* was bought by the private equity firm Recurrent Ventures in 2021.

I grew up reading science magazines. My grandfather subscribed to *Scientific American*, and I pored over the small print and elegant diagrams trying to access the rarefied knowledge I knew they held. I subscribed, myself, to *Discover*, a more accessible but just as expansive and exciting magazine. But, as Sabrina Imbler writes in Defector, "The loss of *Popular Science* also means one less major publication where emerging science journalists can learn the craft and build their careers." Midcareer writers lost an outlet paying decent rates, and new writers lost an internship opportunity. And, a former social media editor at *Popular Science* who was laid off amid the closure told Imbler, scientists lost a prominent venue to have their work featured. "Where are the scientists going to get public investment in their research?"

I grew up reading science magazines because I loved the world and its wonder, but science is much more than awe. Imbler writes,

"Science journalism has arguably never been more important, as the harsh impacts of climate change are hitting the planet faster than many scientists expected and the biodiversity crisis threatens all corners of life on Earth." Climate change, pandemics, reproductive rights—there is no untangling science from the urgent issues in our lives. There is also, though, still the ineffable value of art, of beautiful writing, of knowledge, of joy. All of that, the urgent and the ornamental—the vultures have their claws in it all.

I wish I had a good way to end this foreword. I wish I had a call to action, a rousing plan, a sense of hope. If you're a billionaire, go buy a local paper, steward it with care, trust that the editors who run it know how to do their jobs. There's hope in worker-owned publications, like Imbler's own Defector, an employee-owned publication founded by former Deadspin staffers, or The Sick Times, a newer, scrappier, independent news site covering the Long Covid crisis, founded by two journalists and supported by donations and nonprofit funds. This may be the way forward. This may be the growing pains of a new kind of news industry being born. *Popular Science* was 150 years old, after all. One hundred fifty years ago the telephone and lightbulb were new technologies, and the germ theory of disease was newborn. *Popular Science* was once new, and there are still new things to be made, new ways of telling stories, reaching audiences, and thinking and writing about what science means in the world. But that doesn't make living through what feels like a mass extinction any easier.

JAIME GREEN

And now for a post-script with a whiplash tone shift: If you'd like to submit your work for consideration for a future edition of this anthology, you can find guidelines and a submission form at jaimegreen.net/BASN.

Introduction

THIS VOLUME IS called *The Best American Science and Nature Writing 2024*, because that's the year it comes out, but of course that means the writing is from 2023. I know this is a familiar concept—you've watched the Oscars, after all—and I bring it up simply to emphasize that, in this case, the year really matters: 2023 was the most anomalous year (so far) in human history, the year in which the relationship between people and planet showed its most dramatic signs yet of unraveling.

To review: As the year began, oceanographers started pointing out—with increasing urgency—that sea surface temperatures were higher than they'd ever measured, in pretty much every ocean basin, and by margins that far exceeded what they'd seen before. As the months rolled on, the numbers became increasingly ludicrous, until by late spring they were not just off the chart, but off the wall the chart was tacked to. In June, buoys off the Florida Keys registered ocean temperatures in excess of 101 degrees Fahrenheit, which is what you set your hot tub to. (But you don't keep a coral reef in your hot tub.)

The seas are where most of the planet's extra heat is stored (if they didn't exist, the carbon and methane we've poured into the atmosphere would have already raised the planet's average temperature above 120 Fahrenheit), and as they warm, many things happen. Ice melts, for one, especially at the poles. And when the Arctic Ocean opens, it leads to, again among many other things, earlier snowmelt at higher latitudes. Northern Canada, in other words, dried out a lot sooner than usual in 2023, and then—as land temperatures began to soar—it caught on fire. Caught on fire in ways we've never seen before: vast fires were raging in every province before long. By the time snow finally put them out in late autumn, those fires had produced several times more carbon than all the cooking and heating and cooling and driving and flying done by all the people in Canada, which is what we call a feedback loop. But it had also produced something else equally important: a huge cloud of smoke that drifted south across the border and soon was choking the cities of perhaps the world's most powerful corridor, from Boston and New York south to Washington. As it happens, I was sitting in front of the White House, part of a protest against a gas pipeline, on the day that Washington recorded its worst air quality in recent history; we could barely make out the building we were protesting, though it was a few hundred yards away. It was, I think, useful that the men and women who have allowed America to become the biggest fossil fuel exporter on earth got even a temporary taste of what so many have to deal with day in and day out.

If nothing else, it provided a bit of context for what came next: literally, the hottest day on planet earth in at least 125,000 years. A global chain of satellites and thermometers and ocean buoys offer up a global average temperature each day; the hottest days of the earth year cluster around our summer solstice, because the northern hemisphere has most of the planet's land mass. And this year the numbers beat—by large margins—any day for which we have records. Thermometer data only goes back a couple of centuries, but scientists are masterful at figuring out proxies (ice cores, lake sediments) that extend that record far into paleohistory, which is how we know the true extent of this remarkable stretch of days. They were hotter not than any day in human history, but than any day in much of human prehistory. No members of our species with societies anything like the ones we know have ever lived on an earth so hot.

And, as it turns out, it's not enormous fun living on a world tending toward the hellish. For instance, warm air holds more water vapor than cold, which is my nominee for basic physical fact of the century. That means more evaporation—and hence more drought and fire—in dry places. But once that water is up in the atmosphere, it's going to come down, increasingly in buckets. New

England, where I live, is one of those wet places—we've had a huge increase in gully-washer storms over the past few decades, and this summer in my state of Vermont, it seemed never to stop. Our capital city, Montpelier, was basically closed for months after water surged through most of the town's businesses; there's active talk about somehow relocating the center of the city to higher ground. I live on higher ground, in a tiny town on the spine of the Green Mountains, which got more rain than anywhere else in the course of the summer—a home a mile away was lost in a landslide, and the roads east and west out of town were closed for weeks. And we were lucky—we're a relatively prosperous place. Libya had the biggest rainstorm in its history as summer came to an end, and there the deluge washed away two dams, and then washed away ten thousand people in a single city. They were swept out to sea and drowned.

In case this seems anecdotal to you, the tally at year's end was forty-two weather disasters across the globe that did more than a billion dollars in damage—that's $360 billion dollars total, and less than 40 percent of it covered by insurance. It was the second costliest year on record for drought; flooding in Pakistan did damage equivalent to 9 percent of its GDP. We know that forty thousand people died in Europe from the epic heatwaves that kept sweeping across the continent, but that's because Europe can afford to keep good records of excess mortality. Who knows how many people died because Pakistan had to spend all that money on flood relief and had less left for immunization or education or women's health? Who knows how many people fled their homes for good because there wasn't grass enough left to keep goats alive? We can't count, save by stories, and some of them are in this volume.

We don't know for certain all the reasons the temperature jumped so jaggedly in 2023. An El Niño warming current developed in the Pacific in midyear, and ships kept switching to cleaner fuels, which ironically meant less smoke to reflect the sun's heat back to space. But these paled next to the main trigger: the ongoing accumulation of carbon dioxide, methane, and a few other greenhouse gases in the atmosphere. The project of heating the earth—by far the largest project humans ever contrived—continued its uninterrupted advance. We poured more carbon into the air than ever before, and the concentration of CO_2 in the atmosphere ratcheted past 420 parts per million, another grim milestone. Worse, it felt more than ever as if the most basic physical systems of the planet were starting to strain. If you think about our earth, there are a few overwhelming features: the poles, which are now melting fast (and new data from the Antarctic and Greenland this past year emphasized how much we've underestimated that rate of loss); the great rainforest of South America, where we seem on the brink of savannafication as drought starts to entrench across the Amazon; the boreal forest of the north (which was burning not just in Canada but in Siberia); and the jet stream and the great ocean currents like the Gulf Stream. The jet stream in 2023 was clearly behaving very oddly—driven by the temperature differences between pole and equator, it now gets stuck for long stretches, and oscillates at a greater amplitude, surging heat and cold into unlikely places. And the ocean currents, driven by density differences between saltier and fresher water, seem clearly to be flickering as that same polar ice disappears.

We are rapidly wrecking the earth that has supported us. Not wrecking "the planet," since the third rock from the sun will be there till the sun, in a few billion years, enters its violent dotage. But wrecking the world that we, and all other humans, have known. I want to say once again: In 2023 it got hotter on this earth than it's gotten in at least 125,000 years, roughly the same time as the first evidence of humans etching symbols onto bone. As 2024 began, the researchers were predicting that this year could quite possibly break last year's record.

And yet. And yet. Something else happened in 2023, with the kind of dramatic congruence that strains credulity. In midsummer, in the same stretch of weeks that our instruments were recording those never-seen temperatures, energy analysts were reporting that we had crossed another barrier: we were now installing a gigawatt's worth of solar panels around the world each day. That is to say, we were installing the rough equivalent of a nuclear power plant every single day, but the nuclear power was coming from that giant reactor 93 million miles up in the sky. That's an extraordinary feat—the rapid scaling of this most basic technology for the survival of our species. Solar power has always been seen as "alternative" energy, but now, alongside its sturdy partner Wind, and supported by their companion Battery, the Sun was ready for . . . it's place in the sun? I guess.

This achievement had something to do with policy—half that solar power was going up in China, where governments really do decide these things, and some of the rest was destined for the US, where the hundreds of billions of dollars in President Biden's Inflation Reduction Act had begun to kick in. But it was a reflection of the work done over decades by scientists and engineers, who had managed to make renewable power the cheapest power. Forget alternative energy—this is not the Whole Foods of power, it's the Costco. We live on a planet where the cheapest way to produce energy is to point a sheet of glass at the sun, and that, potentially, changes everything.

It means that if we wanted to we could move with dispatch now. It would not be easy or free, though in economic terms it's mostly gravy: over the past few years, one analysis after another has shown that simply because you'd no longer need to pay for endless shipments of coal and gas and oil, the transition to renewable energy would more than pay for itself. There will be costs in terms of mining: there are few higher tasks right now on planet earth than trying to ensure that the hunt for lithium and cobalt and the rest exacts as small a human and environmental toll as possible, and returns as much as possible to the people and places who will be harmed. But those human and environmental costs do not compare with the damage from continuing to rely on fossil fuel a day longer than necessary. It's not just the existential climate damage I have described (though surely that would be enough); it's also the fact that one death in five on this planet—nine million people a year—stems from breathing the combustion byproducts of fossil fuel. And the fact that relying on energy only available in a few deposits means the people who control them develop wealth and power that usually is abused. Meet the king of Saudi Arabia. Meet the Koch brothers. Meet Vladimir Putin.

Here's a way to picture this change. At the moment, 40 percent of all ship traffic on earth consists solely of shipping coal and oil and gas back and forth around the world to burn. (And the US is the biggest exporter.) That's no longer necessary. Once you've built a solar panel or a wind turbine, the sun and wind deliver the energy for free. Every time the sun rises above the horizon it performs the work of those great ships (and when it sinks below the horizon, a battery can now ensure that the lights stay on). This is a moment of enormous promise—the moment when we can move combustion, with all its dangers, off the surface of the earth and confine it to the heavens. And for cheap.

But that description contains the solution to the riddle of why this simple transition is so hard. It's precisely because the historically richest industry in the world rests on those ships plying the seas, and the pipelines and tanker trucks crisscrossing continents. The fossil fuel industry has prospered like no other for a century simply because you have to pay it every month for a new delivery. If you're Exxon, then the sun delivering your energy for free is the stupidest business model of all time. And if you're Exxon, you have the accrued wealth and power to—well, not to stop this transition. Eventually economics will have its way. But to slow it down. Which is precisely what's happening: we've moved from an era of outright climate denial (save for Donald Trump) to an era of climate delay, when Big Oil works overtime to come up with reasons we should keep burning *something*. They've secured enormous subsidies for schemes like carbon capture, which would allow them to (at great expense, borne by the taxpayer) catch the carbon coming out of their smokestacks and pump it underground. Even if you could make it work, you'd be far better off spending the money building more solar panels. Unless you had a potential fortune in oil and gas reserves that would stay stranded beneath the soil if we did the right thing.

Which is where, and admittedly it's been a winding road, we get to writing. The great question before us as a planet is: Can we make this transition speedily enough to matter? Unlike the other political questions that we face, this one comes with a sharp time limit. If this transition takes too long, then the damage will be unrecoverable: no one really has much of a plan for refreezing the Arctic after it's melted. And so the fight of activists—and I am one—is to try and catalyze this transition so it happens faster. The job of writers—and I am one of those as well—is not quite that. It is to tell the story of our moment as honestly and vividly and beautifully as possible, so as to reach our hearts, our imaginations, our self-interest. I am confident that if we manage that then we will prevail.

Or maybe *confident* is the wrong word. We don't know. I fear I've written more words about global warming than anyone in the English language, and by a considerable margin—which, given the state of the thermometer, might make me the least successful writer in history. But as I say, in 2023, you could sense the tide beginning to turn; the question now is how fast that tide will run. If it races we have a chance.

If this seems too political a role for artists to play—and doubtless even for many writers included in this volume that would be the case—I offer only this. One job of every art form is to help preserve itself. Which is to say, I hope very much that there is a *Best American Science and Nature Writing 2124*, though I will not be here to enjoy it.

In the roughest terms, this collection is divided into two sections. The first is about the problem itself, and the second about the world on which it is taking place. If you can work your way through the first, you will be rewarded with the second—with, above all, a reminder that this world is still a lovely and deep place, well worth the fighting for. (More than most years, the nature follows the science.)

My schema isn't perfect, and it's also unfair. There were myriad wonderful pieces of writing in the last year that I couldn't shoehorn into my vision for this volume, and so they have been cast aside. They include echoes of the Covid pandemic (which dominated this book in recent years) and foreshadowings of the AI revolution, and a dozen other topics fascinating and important. But I knew what I was getting into—some years ago I edited, for the Library of America, a volume called *American Earth* that has become one of the standard collections of environmental literature. That experience taught me that the job of the anthologist is woeful, discarding one gem after another, always with the bitter sense of the injustice one is performing. In my defense I offer only the excuse that we are living in extraordinary times—indeed, as I have said, that 2023 may be the most extraordinary of years. Until, perhaps, 2024.

At any rate, we begin with a few pieces from the newspaper. Magazine essays tend to dominate these volumes, and for good reasons—that's usually where one has the room to stretch out and do some real reporting and writing, less constrained by space and by the conventions of style that dominate the broadsheets. But one of the remarkable developments of the last few years—spurred, I think, by the rise of Greta Thunberg and the youth climate movement a few years ago—is that the great metropolitan newspapers (which in the American case at least means the *New*

York Times, the *Washington Post,* and the *Los Angeles Times*) built dedicated climate desks with superb reporters, and with backup from savvy practitioners of the new art of data visualization. Time and again they have produced reports with those cascades of pictures or illustrations of statistics come to life—your fire risk by zip code, say. We can't re-create those in these more old-school static pages, but I wanted to give some sense of just how professional, comprehensive, and depth-defined this reporting has become. And it's also unafraid, no longer as bound by the devotion to false balance that for two decades developed into a he-said/she-said story that left readers less sure than they should have been of the state of the science. So the two reporting pieces from the *Times* and the *Post,* and a perceptive and gutsy column from the invaluable *LA Times* climate analyst Sammy Roth, give you a sense of how this story is being covered by those who have to deal with it daily.

Even if the basic science behind the climate crisis has been well understood for nearly four decades (and in some ways for the better part of three centuries), remarkable research continues, much of it in places that make the work crushingly hard. Consider, for instance, Sarah Kaplan's account of drilling ice cores up on the Greenland ice shelf, or Douglas Fox's chronicle of how scientists are coming to understand the complex biology of the Amazon rainforest. (Think vines, not trees.) There are only a few physical features on this planet so large that their failure affects the globe; it's no wonder, then, that the Amazon is also the setting for Alex Cuadros's much larger look at what might happen as that lush jungle turns to savanna. And this seems like the right place for Amanda Gefter's look at what we're coming to understand about plant-based intelligence.

What to do about climate change is conceptually simple—above all, replace fossil fuels with renewable energy—but this is the largest single economic transition humanity has ever tried: basically we're rebuilding the engine while the plane is in the air. If renewable energy is cleaner than coal, gas, and oil, that doesn't mean it's without problems, and Nick Bowlin gives a thorough tour of the pitfalls around lithium, a crucial green mineral. This transition will also require so much cash that it attracts hustles and grifts of all kinds. Heidi Blake offered *New Yorker* readers a classic account of carbon offsets and how they grew into a global scandal.

In the end, though, climate change is most likely to reach us through the stories of human tragedy that now arrive almost daily. Veteran climate writer Carolyn Kormann offers a particularly nuanced account of the Lahaina wildfire tragedy.

It was a brutal year on planet earth. But, of course, it was also a glorious, beautiful, buzzing, mysterious, wonderful year on planet earth, like every other year. Because even though we're doing huge damage, there remains vital life bursting out all over. I want to mark the transition to this part of the book with a lovely essay from Ian Frazier. When I was a young writer a week out of college, I shared an office at the *New Yorker* with Mr. Frazier, who was already a legend within the magazine—the writer's writer. This piece—about the largest beaver dam on earth—was on the Yale E360 website, and if it doesn't make you grin then you need to read it twice.

Ben Goldfarb—another beaverphile, given his award-winning book *Eager*—has switched to birds this year, and his account of Chicagoans working to prevent birds from crashing into the city's expanse of glass buildings is a good reminder that, while we struggle with the very large question of climate change, there are some things closer to hand that are easier to fix. (Easier, not easy.)

Douglas Fox, this time writing in *Science News*, brings us an account of the astonishing life being discovered beneath Antarctic glaciers (a reminder of the maxim that the least hospitable inch of our earth is a hundred times *more* hospitable to life than anywhere else we know of out there in the universe).

Along somewhat the same lines, Brendan Egan reports on the nature that exists even in the "total shit land" behind his Midland Texas home. "A chunk of it was irrigated for grazing a herd of maybe a dozen cattle, there were high tension powerlines, a submerged gas pipeline, and some lots slated to be fracked and drilled, but mostly it was mesquite, bunchgrass, and plastic bags." And yet.

The poet B. M. Owens describes another desolate place—the inside of the aquarium where a poor whale lives out a "life."

Dan Musgrave takes us inside the mind of another animal—a bonobo named Nathan—in a long and satisfyingly complicated account of interlocking lives.

Reading each of those made me think immediately of Elizabeth Kolbert's essay in the *New Yorker* about whether or not we will be able to use AI to speak across the boundaries of species. And Isobel

Whitcomb offers a marvelous chronicle of the efforts of the Klamath tribes to protect the suckerfish of the Pacific Northwest, while Joe Spring takes us to the largest protected area in the country, though with precious little land to support the animals that try to call it home. It's all complicated, and all beautiful, as Emma Marris reminds us in her account of the eagles returned to Scotland and the complications they cause.

We're getting to the end now, and the underlying theme of this volume, more clearly glimpsed, is life and death—the attempts to stave off the death of a lively planet. But death is natural, too, or at least it can be—Lindsay Ryan wrestles with the conundrum in an account of her mother's decision to use her state's Death with Dignity Act.

Rachel May offers some notes on the relation between the natural world and our own sanity in her recollections of a tree at one of the country's most famous mental institutions.

But I wanted to leave the final word to a scientist intimately involved in the fight to avert the tragedies we face. Ayana Elizabeth Johnson, the marine biologist and activist, now living in down east Maine, offered the commencement address at my college, Middlebury, in the spring of 2023, and a week later *Time* magazine reprinted her address in full—a wise choice, I think, as it lays out a serious case for a certain kind of hope.

"To address the climate crisis, the all-encompassing challenge that will touch whatever life and work you will go on to, requires that we not just change or adapt, but that we *transform* society, from extractive to regenerative," she said. "This is a monumental task. And it requires that we focus not on endless analysis of the problem, but on summoning an expansive sense of *possibility*, on harnessing our imaginations and our creativity." That, I think, could be the epigraph for this volume, and maybe for every volume in this charmed series.

BILL McKIBBEN

The Best
AMERICAN
SCIENCE &
NATURE
WRITING
2024

ANNIE GOWEN

Climate-Linked Ills
Threaten Humanity

FROM *Washington Post*

THE FLOODS CAME, and then the sickness.

Muhammad Yaqoob stood on his concrete porch and watched the black, angry water swirl around the acacia trees and rush toward his village last September, the deluge making a sound that was like nothing he had ever heard. "It was like thousands of snakes sighing all at once," he recalled.

At first, he thought villagers' impromptu sandbags, made from rice and fertilizer sacks, had helped save their homes and escape Pakistan's worst floods on record. But Yaqoob—whom villagers call a wadero, or chief—soon realized it was just the beginning of a health disaster. The temperatures rose to triple digits, as the water that would not recede festered in the sun.

An elderly woman died in a boat on the way to the hospital, overcome by heat and dehydration. Dark clouds of mosquitoes bit through even the toughest donkey's hide, spreading malaria to Yaqoob and four dozen of his neighbors. People came down with itchy dermatitis from walking through the floodwaters. Farmers who could not plant in drenched fields began cutting back their simple meals of vegetables and rice from three a day to two. And then, for some, just one.

"I had no idea what miseries this flood would bring for us," said Yaqoob, whose village is in Sindh, the hardest hit province in a disaster that left a third of the country underwater.

Pakistan is the epicenter of a new global wave of disease and death linked to climate change, according to a *Washington Post* analysis of climate data, leading scientific studies, interviews with experts and reporting from some of the places bearing the brunt of Earth's heating. This examination of climate-fueled illnesses— tied to hotter temperatures, and swifter passage of pathogens and toxins—shows how countries across the globe are ill-prepared for the insidious, intensifying risks to almost every facet of human health.

To document one of the most widespread threats—extreme heat—the *Post* and CarbonPlan, a nonprofit that develops publicly available climate data, used new models and massive data sets to produce the most up-to-date predictions of how often people in nearly 15,500 cities would face such intense heat that they could quickly become ill—in the near-term and over the coming decades. The analysis is based on a measure called wet-bulb globe temperature (WBGT), which takes into account air temperature, humidity, radiation and wind speed, and is increasingly used by scientists to determine how heat stresses the human body.

The *Post* analysis showed that by 2030, 500 million people around the world, particularly in places such as South Asia and the Middle East, would be exposed to such extreme heat for at least a month—even if they can get out of the sun. The largest population—270 million—was in India, followed by nearly 190 million in Pakistan, 34 million across the Arabian Peninsula and more than 1 million apiece in Mexico and Sudan.

The results show how the risk has been growing and will escalate into the future. The number of people exposed to a month of highly dangerous heat, even in the shade, will be four times higher in 2030 than at the turn of the millennium.

By 2050, the number of people suffering from a month of inescapable heat could further grow to a staggering 1.3 billion. At this point, vast swaths of the Indian subcontinent will swelter under extreme humid heat, as will parts of Bangladesh and Vietnam. Only those who can find cooling will find respite.

To reach these estimates, the *Post* and CarbonPlan combined one of the most detailed sets of historic heat data with the latest climate projections produced by NASA supercomputers, offering one of the most detailed estimates of current and future heat stress at a local level ever produced. The projections assume countries make steady progress toward cutting planet-warming emissions, as they have committed to do.

The *Post* defined its dangerous heat threshold as more than 89.6 degrees Fahrenheit wet-bulb globe temperature, equal to a temperature of 120 degrees on a dry day, or mid-90s temperature on a very humid day. Spending more than fifteen or so minutes beyond that limit, many researchers say, exacts a harsh toll on even a healthy adult; many deaths have occurred at much lower levels.

Extreme heat, which causes heatstroke and damages the heart and kidneys, is just one of the ways that climate change threatens to cause illness or kill.

So far this year, more than 235,000 Peruvians have come down with dengue fever and at least 399 died, according to Peru's national center for disease control, the most in that nation's history. Smoke from record-breaking Canadian wildfires billowed across the United States, triggering asthma attacks that forced hundreds to seek hospital care. And East Africa's worst drought in at least forty years, which has spurred widespread risk of famine, is 100 times more likely to have happened because of human-caused warming, researchers say.

The number of heat-related deaths of people over sixty-five increased by 68 percent from 2017 and 2021 compared with between 2000 and 2004, according to a peer-reviewed report from the *Lancet* last year, while the months of favorable conditions for malaria in the Americas' highlands rose by 31 percent between 2012 and 2021 compared with sixty years earlier.

"We can say now that people are dying from climate change, and that's a different kind of statement than we would have made before," said Kristie L. Ebi, a professor in the Center for Health and the Global Environment at the University of Washington who co-authored the 2022 Lancet Countdown report. "Climate change is not a distant threat to health, it's a current threat to health."

Many of the most affected countries have contributed the least to the climate crisis, and are ill-prepared to manage the rapidly multiplying threats.

Last year in Pakistan, dangers piled one atop the other. First, the country suffered a record-breaking heat wave beginning in March. Fires rampaged through its forests. Record high temperatures melted glaciers faster than normal, triggering flash floods. And then heavy monsoon rains caused unprecedented floods, which left 1,700 dead, swept away 2 million homes and destroyed 13 percent of the country's health-care system.

"Pakistan's crisis was almost prophetic," said Sherry Rehman, Pakistan's outgoing climate minister, in a phone interview. "Look at this summer." As the world shatters temperature records—this year is now likely to be the warmest in recorded history—she said, "countries like us, the hot spots, are going to feel the burn immediately."

Inside the Ward

On a recent 109-degree day, babies wailed and adults vomited into buckets in the crowded heatstroke ward of Syed Abdullah Shah Institute of Medical Sciences, a 350-bed government medical center in central Sindh. With just seven beds for heatstroke victims, patients' parents and relatives crowded together on the mattresses. Nurses in green scrubs attached bags of intravenous hydration fluids to the arms of even the tiniest patients as fans whirled and two air conditioners dripped and chugged.

The number of heatstroke patients coming to the hospital in summer has increased around 20 percent a year in the last five years, according to M. Moinuddin Siddiqui, the hospital's medical director, at a time when Pakistan experienced three of its five hottest years on record.

The changing climate has affected people in painful ways, Siddiqui said, including high-grade fevers, vomiting, diarrhea and related diseases such as gastroenteritis. "I have been a doctor here for two decades and such climate changes I have not seen before. It's disheartening," he said.

The proliferation of climate ills has taxed this regional hospital center at the same time it has taken in patients from twelve nearby clinics and medical dispensaries swept away in the flood, he said.

The hospital has taken a variety of "special measures" to support the heat patients, including creating the small stroke unit, where patients are treated before either being admitted or sent home with electrolyte powder packets for rehydration. They also added air conditioners in every ward, but sometimes even those don't cool enough to make patients comfortable.

Despite such preparations, he said, last year's heat wave shocked the whole system. The air conditioning shut down under intense use and a huge crowd amassed inside the hospital, creating a "panic-like situation" for both the patients as well as health-care providers.

Farm laborers are routinely brought in unconscious with high fevers and may even end up on a ventilator, the doctor said. Outdoor workers are at increasing risk of heat-related illness, but their low-wage jobs are a lifeline. About half of Sindh's population lives in rural areas, according to a World Bank report, and 37 percent of that population lives below the poverty line.

Siddiqui finds it difficult to tell them to avoid working in the oppressive heat when they earn the equivalent of just a few dollars a day.

"If they take rest in the house they go hungry!" he said.

Around Sindh, women and child specialists and nurses say that they are seeing a rise in miscarriages, low-birth-weight babies and decreased production of breast milk—that they blame on the stress from the floods, along with rising summer temperatures.

"Miscarriages have been increasing because of the intense heat," said Zainab Hingoro, a local health-care worker. When she once would have three out of ten pregnant patients miscarry, she now has five to six out of ten. The number of low-birth-weight babies is "drastically increasing," she added.

Sughra Bibi, thirty-eight, who was about to deliver her sixth child, said she suffered frequent kidney pain and gastrointestinal upset from drinking unsafe water.

"I am not well," she said, adding that her husband, a laborer, struggled to get enough food to sustain her pregnancy. The couple still lives in a temporary tent nearly a year after the floods, and she wept as she showed photos of her children, ages nine and six, who died in the floods' chaotic aftermath.

Insect-borne diseases are also on the rise. Siddiqui said his hospital saw a "very unusual" influx of malaria patients during February through June, a time not generally considered peak malaria season.

After the floods, Pakistan grappled with over 3 million suspected malaria cases, up from 2.6 million in 2021, according to the World Health Organization. The outbreak was spurred on by standing water and other circumstances making it easier for mosquitoes to breed, reversing decades of progress of reducing cases.

Malaria kills more than 600,000 people a year around the world, and studies show that climate change is driving the once tropical disease to higher altitudes and new areas. A study last year by Pakistan's Global Climate-Change Impact Studies Center showed that dengue—another mosquito-borne illness—will begin appearing in far higher altitudes by the end of this decade.

Yaqoob, sixty-two, the chief of Bagh Yusuf village, has made two trips to the hospital's malaria ward in the past year.

Yaqoob, a retired primary schoolteacher, moves around Bagh Yusuf's rocky lanes on crutches, after he lost a leg a decade ago when bandits shot him while trying to steal his scooter.

The village of concrete and thatched roof dwellings sits up on a dune, so people there can catch cooling breezes in the summer. On hot nights, they sleep outside on string cots, called charpoys, covered in the colorful quilts the region is known for. Still, the heat can be brutal. Villagers drink a combination of jaggery—sugar cane—and black pepper water they say wards off heatstroke.

When the floodwaters lapped at their doorsteps last fall, villagers kept them at bay and Yaqoob held out his crutch to help save several people from drowning.

But three months of living surrounded by contaminated water that smelled like the corpses of dead animals took its toll. First, one of his neighbors sparked a fever, then another. Getting sick in Bagh Yusuf was never easy, even before the flood. Villagers go to a small dispensary if they fall ill: A private doctor costs too much and a trip to the hospital is a last resort.

After fifteen days, it was Yaqoob's turn.

He was overcome with a fever stronger than he had ever experienced and began bleeding from his nose. Relatives had to take him out by boat to the hospital.

Once there, he remained unconscious most of the time. "I hallucinated that the water had reached my house and I had to keep my family members safe. Another time, I thought my siblings were in bad condition and living in a roadside shelter," he said.

He recovered after about a week, but relapsed in July, spending two more days in the hospital before doctors said he was strong enough to go home.

"Hotter and Hotter"

One June afternoon, a bread maker in Jacobabad, Pakistan—which has temperature highs in the summer months so extreme it's often called "the hottest place on Earth"—sat outside when it was 111 degrees, flipping rounds of dough into the air and toasting them over hot coals.

The air around his workspace outside a downtown restaurant is always several degrees hotter than the normal air temperature, Dil Murad said, which can often be overwhelming. He said he feels trapped in his job as the summer heat intensifies, and tries to keep as cool as he can by drinking large amounts of water every hour.

"It's difficult because this scorching heat has become unbearable," said Murad, twenty-five. "I don't have any other source of income, and I have to feed my kids. It's the only craft I know."

During a devastating heat wave last year that lasted weeks and vented misery across Pakistan and India, the temperature in Jacobabad soared to a world high of 123.8 degrees on May 14. Human-caused climate change made this record-breaking heat wave at least thirty times more likely, according to modelers at the World Weather Attribution initiative. About fifty people died in Jacobabad alone, according to one estimate.

When temperatures soar life slows to a near halt in this city of 170,000, where the streets are crowded with men wearing loose white cotton clothes and women in headscarves who jostle for space with farmers driving donkey carts. Residents who can't afford air conditioning try to not move and stay indoors or search for a patch of shade. Sometimes their only respite is a slow-moving fan run by a single solar panel—which only works during the day.

Sweaty rickshaw drivers and construction workers crowd around volunteers passing out cooling herbal drinks made of the bluish-red falsa berries, and residents buy blocks of ice from the area's busy ice factories to keep themselves—and their food—cool.

In villages outside the city, farmworkers still venture into the rice, wheat and fodder fields, but try to rest during the hottest part of the

day, from noon until about 3 p.m. Even then, some become dizzy
and collapse. Cows and buffalo—their ribs visible—take refuge in
ponds.

The number of days when Jacobabad's temperature surpassed
113 degrees rose from 12 between 2011 and 2015 to 32 between
2016 to 2020, according to an analysis by Aga Khan University.

"It has gotten hotter and hotter," said Muhammad Yousif Shaikh,
the deputy commissioner for the Jacobabad District. "For some vulnerable communities, the weather has become simply unbearable."

Shaikh said the district is working to put in place long-term
solutions to rising temperatures, such as shoring up the community's shaky water infrastructure and planting shade trees lost to
unplanned development.

But residents have said that they have done little to help them
during the hottest days. The district had no permanent heatstroke
center until the height of the heat wave last May, when a local
NGO, the Community Development Foundation, helped establish
one in a local hospital. It has only eight beds.

"During last year's heat the government did not do anything for
us, not even water, nothing," said Mukhtiar Bhatti, the head of Pir
Bux Bhatti, a village about 11 miles north of Jacobabad.

Researchers who have formed a group dubbed the Climate Impact Lab found in a recent study that heat-related mortality will
expand dramatically in the coming decades and in the world's
poorest and hottest places, exacerbating inequality.

They projected that higher temperatures will lead to a staggering 150,000 added deaths per year in Pakistan by 2040—unless
the country can grow substantially more wealthy and better adapt
to frequent bouts of extreme heat. The rising death rate, 50 per
100,000, is higher than that of nearly all other countries, barring
some of the least developed parts of Africa and the Middle East. It
is more than twice the number estimated for neighboring India,
which has more financial resources to shield its population from
the worst climate impacts.

"The way the rich countries are going to respond is by spending
more to protect ourselves, and in many parts of the world those
opportunities don't exist," said Michael Greenstone, a University
of Chicago economist and the study's co-author.

In many cases, improving odds of survival means one thing—
access to air conditioning. A study led by scientists at the University

of California at Berkeley projects that less than one in ten Pakistani households will have air conditioning in 2030, compared with 25 percent of Indian homes. In the United States, 92 percent of residents had air-conditioned homes as of 2021.

Jacobabad has always had high temperatures in the summer but climate change is fueling heat waves that arrive earlier and last longer than ever before, which may eventually make the area uninhabitable for even healthy humans, experts say. In Jacobabad, a wet-bulb globe temperature of at least 90 degrees will prevail for a third of the year by 2030, the *Post* analysis found.

Scientists say the higher the wet-bulb globe temperature climbs, the more difficult it becomes to keep cool and the heart and the kidneys can fail as they work overtime to maintain blood pressure and the flow of fluid in the body.

"As the temperature begins to rise, in order to lose enough heat, you have to sweat," said Zac Schlader, an associate professor at Indiana University at Bloomington who studies the physiological impact of extreme heat. "And that evaporation of that sweat is dependent on the amount of water vapor that's in the air."

If the air is too moist to absorb sweat, a person's internal body temperature will continue to rise. The heart pumps faster and blood vessels expand to move more blood closer to the skin, in order to cool off. At the same time, the brain sends a signal to send less blood to the kidneys to stop losing liquid through urine, which deprives the kidneys of oxygen.

The wet-bulb globe temperature combines the regular air temperature ("dry-bulb"), the humidity-adjusted temperature ("wet-bulb") and the radiant heat from the sun and hot surfaces ("globe temperature") to capture heat stress.

While every human body is different, many experts and institutions cite just under 90 degrees as the wet-bulb globe temperature beyond which the risks of heat illness become very severe. The U.S. Marine Corps cancels all physical training at 90 degrees. The National Weather Service says that in much of the United States, that threshold represents an "extreme threat" to health and it will stress the body after working in direct sunlight for just 15 minutes. A study in Taiwan found that on days reaching a wet-bulb globe temperature of at least 89.6 Fahrenheit (32 Celsius), heat-related emergency hospital visits increased by about 50 percent compared to other warm season days.

Even lower temperatures pose "a very real risk to human health," Schlader said, especially for vulnerable people.

Temperatures had reached 122 degrees one day in Pir Bux Bhatti during last year's heat wave when Fazeela Mumtaz Bhatti, forty-six, rose to prepare breakfast for her husband, Mumtaz Ali, fifty, and their eleven children.

Bhatti—who was otherwise healthy—had made a bit of potato and charred bread, working in a poorly ventilated brick room on an open fire fueled by dung patties. Around 1 p.m., she began to complain she wasn't feeling well, her daughter Naheed, eighteen, recalled.

Bhatti left the house to walk a few dozen yards and collapsed, face first, in the dust. In a panic, Naheed ran to help, cradling her mother's head in her arms and trying to ply her with water combined with sugar and salt to help her rehydrate. Other women in the village rushed to assist, moving the woman back into the small house and onto a string cot, where they doused her body with water from a nearby pump and tried to keep her calm.

"She was fire to the touch," Naheed recalled. "She just kept saying, 'Don't you worry about anything, I'll be okay. Just make sure your father and siblings are fed.'"

But Bhatti's condition worsened, and her husband raced to borrow a car to take her to the hospital in the city, some distance away. By the time they reached the hospital, she was already dead.

Naheed mourns the loss of her mother, who often spoke of finding Naheed a good man to marry, and used to tease her eldest daughter by saying, "You are only a guest here, you only have so much time to live in your father's house." In quieter moments, she would tell Naheed, "You have to find courage within yourself because life is difficult."

Now, Naheed is left to manage the housework and care for the large family on her own.

"We just couldn't keep her safe and alive," she said quietly. "It's difficult for me, but I have to take care of my brothers and sisters. I just try and cope with it."

"People Have Forgotten Us"

In Bagh Yusuf, life has returned to some semblance of normalcy after the floods, but several aftershocks remain. All but about six of the families who had fled returned. The residents were able to clear the cemetery and have their annual religious festival, where they pray to their ancestors and celebrate with a mutton feast. The farmers who live in their village revved up their gaily decorated red tractors and began planting again.

But hunger remains a problem.

Muhammad Ishaq, forty-two, lost his cotton crop during the flood, along with the $81 he'd invested in seed and insecticide. After the floods, the debt made farming impossible, so he began laboring as a stone crusher for about $3 a day. In April, he was able to sow his cotton crop, he said, but water is scarce.

"We hardly eat two times a day," he said. They generally eat bread, okra or potatoes for breakfast, lentils for lunch and goat milk and bread for dinner. The younger of his five children often whimper and cry from hunger, he said.

His oldest son, Tariq, seventeen, has been working in construction which has allowed them to buy more food. But it also puts him more at risk, because he'll be laboring outdoors.

Pakistan—a fast-growing country of 241 million—had myriad challenges even before the floods, with a high percentage of poverty, low literacy rate, vanishing water supply, rising inflation and ongoing political turmoil after last year's ouster of former prime minister Imran Khan, now jailed, with elections set for the coming months.

Officials in Pakistan say that the scale of the flood disaster was so epic—"biblical" in the words of Rehman—that it was beyond their ability to respond, with total damage to the economy estimated at $30 billion. They say they now need $13 billion in additional international support—on top of $16 billion already pledged—to prepare their country for future disasters.

Pakistan wants to use the additional money to expand its network of hospitals in rural areas, move residents out of flood plains and bolster its water supply. The government of Sindh is already working with the World Bank to replace lost mud brick dwellings with 350,000 homes that will have rainwater harvesting systems and latrines.

Ali Tauqeer Sheikh, a climate change specialist based in Islamabad, said the country needs to upgrade construction standards to withstand more extreme weather, shore up its reserves of emergency food and water for the next crisis and develop heat action

plans for its cities and provinces. The only city with a significant plan
to address a heat wave emergency is Karachi, with one he helped
write after a deadly heat wave there killed more than 1,200 in 2015.

Muhammad Jaohar Khan, a health specialist with UNICEF
in Islamabad, said that the floods—which submerged more than
2,000 health-care facilities—ratcheted up pressure on a system
that was already burdened and failing to reach the poor in rural
provinces like Sindh. Even before the floods, poor nutrition had
stunted the growth of 40 percent of the children under five in
Pakistan.

"These districts were already deprived, and had been hit several
times by floods and droughts," he said. "They went from the bad
to worse category."

Samuel S. Myers, a principal research scientist at the Harvard
T.H. Chan School of Public Health, says that one of the biggest
threats South Asia faces is malnutrition, as climate change harms
crops. The global rate is rising again after years of declines, with
more than 800 million people at risk for malnutrition in 2022.

Children are among the most vulnerable to rising temperatures, which affect pregnant women and disrupt food production.

At the Jacobabad Institute of Medical Sciences, Kamala Bakht,
a doctor in the infant nutrition center, said that the number of
low-birth-weight babies entering the feeding program had been
steadily increasing since 2018—from about forty to fifty-five a
month.

She says more intense heat—which exacerbates dehydration,
putting mothers at risk for miscarriages—has played a role, as well
as the floods, which had a "great impact" on her patients and their
ability to properly nourish themselves and their newborns.

Inside one of the feeding rooms, a woman named Pathani cradled her tiny son, Allah Dino. She had worked for three of her
earlier pregnancies, she said, harvesting rice in the heat, and had
miscarried each time. With this latest pregnancy, she had stayed
indoors, but then came down with typhoid and delivered the baby
prematurely—at eight months. When she first arrived at the feeding center, Allah Dino weighed 2.4 pounds, she said. Ten days
later, his weight was 2.6 pounds.

If Pathani's son lives to be twenty-seven years old, at that point
Pakistan will experience more than two months of highly dangerous heat each year, even in the shade.

After years of resistance by richer nations, Pakistan and other developing nations also pushed through a breakthrough "loss and damage" fund at global climate talks last year, where richer countries like the United States—which have contributed the bulk of the world's greenhouse gas emissions—will give money to poorer nations bearing the brunt of the impacts.

In the coming months, as countries gather for the UN Climate Change Conference in Dubai, delegates will push wealthy nations to spell out how the loss and damage fund will work.

But Rehman says that with so many countries now facing their own climate emergencies, they will be less likely to want to help.

"Already I hear ministers saying we need to spend money in our country now," she said. "People have forgotten us."

Originally published in the *Washington Post* with photographs by Saiyna Bashir and data visualization by Niko Kommenda

RAYMOND ZHONG

The Grand Canyon, a Cathedral to Time, Is Losing Its River

FROM *New York Times*

DOWN BENEATH THE tourist lodges and shops selling keychains and incense, past windswept arroyos and brown valleys speckled with agave, juniper and sagebrush, the rocks of the Grand Canyon seem untethered from time. The oldest ones date back 1.8 billion years, not just eons before humans laid eyes on them, but eons before evolution endowed any organism on this planet with eyes.

Spend long enough in the canyon, and you might start feeling a little unmoored from time yourself.

The immense walls form a kind of cocoon, sealing you off from the modern world, with its cell signal and light pollution and disappointments. They draw your eyes relentlessly upward, as in a cathedral.

You might think you are seeing all the way to the top. But up and above are more walls, and above them even more, out of sight except for the occasional glimpse. For the canyon is not just deep. It is broad, too—18 miles, rim to rim, at its widest. This is no mere cathedral of stone. It is a kingdom: sprawling, self-contained, an alternate reality existing magnificently outside of our own.

And yet, the Grand Canyon remains yoked to the present in one key respect. The Colorado River, whose wild energy incised the canyon over millions of years, is in crisis.

As the planet warms, low snow is starving the river at its headwaters in the Rockies, and higher temperatures are pilfering more of it through evaporation. The seven states that draw on the river are using just about every drop it can provide, and while a wet winter and a recent deal between states have staved off its collapse for now, its long-term health remains in deep doubt.

Our species' mass migration to the West was premised on the belief that money, engineering and frontier pluck could sustain civilization in a pitilessly dry place. More and more, that belief looks as wispy as a dream.

The Colorado flows so far beneath the Grand Canyon's rim that many of the four million people who visit the national park each year see it only as a faint thread, glinting in the distance. But the river's fate matters profoundly for the 280-mile-long canyon and the way future generations will experience it. Our subjugation of the Colorado has already set in motion sweeping shifts to the canyon's ecosystems and landscapes—shifts that a group of scientists and graduate students from the University of California, Davis, recently set out to see by raft: a slow trip through deep time, at a moment when Earth's clock seems to be speeding up.

John Weisheit, who helps lead the conservation group Living Rivers, has been rafting on the Colorado for over four decades. Seeing how much the canyon has changed, just in his lifetime, makes him "hugely depressed," he said. "You know how you feel like when you go to the cemetery? That's how I feel."

Still, every year or so, he comes. "Because you need to see an old friend."

The lands of western North America know well of nature's cycles of birth and growth and destruction. Eras and epochs ago, this place was a tropical sea, with tentacled, snail-like creatures stalking prey beneath its waves. Then it was a vast sandy desert. Then a sea once again.

At some point, energy from deep inside the Earth started thrusting a huge section of crust skyward and into the path of ancient rivers that crisscrossed the terrain. For tens of millions of years, the crust pushed up and the rivers rolled down, grinding away at the landscape, up, down, up, down. A chasm was cleaved open, which the meandering water joined over time with other canyons, making one. Weather, gravity and plate tectonics warped and sculpted the exposed layers of surrounding stone into fluid, fantastical forms.

The Grand Canyon is a planetary spectacle like none other—one that also happens to host a river that 40 million people rely on for water and power. And the event that crystallized this odd, uneasy duality—that changed nearly everything for the canyon—feels almost small compared with all the geologic upheavals that took place before it: the pouring, 15 miles upstream, of a wall of concrete.

Since 1963, the Glen Canyon Dam has been backing up the Colorado for nearly 200 miles, in the form of America's second-largest reservoir, Lake Powell. Engineers constantly evaluate water and electricity needs to decide how much of the river to let through the dam's works and out the other end, first into the Grand Canyon, then into Lake Mead and, eventually, into fields and homes in Arizona, California, Nevada and Mexico.

The dam processes the Colorado's mercurial flows—a trickle one year and a roaring, spiteful surge the next—into something less extreme on both ends. But for the canyon, regulating the river has come with big environmental costs. And, as the water keeps dwindling, plundered by drought and overuse, these costs could rise.

As recently as a few months ago, the water in Lake Powell was so low that there almost wasn't enough to turn the dam's turbines. If it fell past that level in the coming years—and there is every indication that it could—power generation would cease, and the only way water would be released from the dam is through four pipes that sit closer to the bottom of the lake. As the reservoir declined further, the amount of pressure pushing water through these pipes would diminish, meaning smaller and smaller amounts could be discharged out the other end.

If the water dropped much more beyond that, the pipes would begin sucking air, and in time Powell would be at "dead pool": Not a drop would pass through the dam until and unless the water reached the pipes again.

With these doubts about the Colorado's future in mind, the UC Davis scientists rigged up electric-blue inflatable rafts on a cool spring morning. Slate-gray sky, low clouds. Cowboy coffee on a propane burner. At Mile 0 of the Grand Canyon, the river is running at around 7,000 cubic feet per second, rising toward 9,000—not the lowest flows on record, but far from the highest.

Cubic feet per second can be a little abstract. As the group paddles toward the canyon's first rapids, Daniel Ostrowski, a master's student in agronomy at Davis, says it helps to think of basketballs. Lots of them. A regulation basketball fits loosely inside a foot-wide cube. Draw a line across the canyon, and imagine 9,000 basketballs tumbling past it every second.

At Mile 10, the scientists float by a more tangible visual aid. Ages ago, a giant slab of sandstone plunged into the riverbed from the cliffs above, and now it looms over the water like a hulking Cubist elephant. Or at 9,000 basketballs per second it looms. At higher flows—12,600 basketballs, say—it's submerged to its knees. At three times that, the water comes up to its head. And at 84,000, which is how much ran through in July 1983, the elephant is all but invisible, a ripple at the river's surface.

The big problem with low water in the canyon, the one that compounds all others, is that things stop moving. The Colorado is a sort of circulatory system. Its flows carved the canyon but also sustain it, making it amenable to plants, wildlife and boaters. To understand what's happened since the dam started regulating the river, first consider the smallest things that its water moves, or fails to move.

The Colorado picks up immense amounts of sand and silt charging down the Rockies, but the dam stops basically all of it from continuing into the Grand Canyon. Downstream tributaries, including the Paria and Little Colorado, add some sediment to the river, but not nearly as much as gets trapped in Lake Powell. Plus, when river flows are weak, more sediment settles on the riverbed.

The result is that the canyon's sandy beaches, where animals live and boaters camp at night, are shrinking. Beaches that were once as wide as freeways are today more like two-lane roads. Others are even scrawnier. The sandy space that remains is also becoming overgrown with vegetation: cattail and brittlebush, arrowweed and seepwillow, bushy tamarisk and spiny camelthorn. Before the dam came in, the river's springtime floods regularly washed this greenery away.

A lusher, less-barren canyon might not sound like a bad thing. But grasses and shrubs block the wind from blowing sand onto the slopes and terraces, where hundreds of cultural sites preserve the history of the peoples who lived in and around the canyon. Sand shields these sites, which include stone structures, slab-lined granaries and craterlike roasting pits, from weather and the elements. With less sand drifting up from the riverside, the sites are more exposed to erosion and trampling by visitors.

Also, not every place in the canyon is becoming greener. Drought can sap the water that courses within the porous stone walls, water that, where it spurts out, sometimes feeds eye-popping bursts of plant life. Lately, some of these springs, like Vasey's Paradise at Mile 32, have dried to a dribble for long stretches. But a few bends downriver, the UC Davis scientists spot several hanging gardens that, for now, are still thriving.

Besides sand, the Colorado is failing to move larger objects in the canyon. Cobbles and boulders periodically tumble in from hundreds of tributaries and side canyons, often during flash floods, creating bends and rapids in the river. With fewer strong flows to whisk this debris away, more of it is piling up at those bends and rapids. This has made many rapids steeper and narrowed boaters' paths for navigating them.

Today, when the water is low, more boulders in the river are exposed at certain rapids, making them trickier to negotiate for the 30-to-40-foot-long motor rigs that are popular for canyon tours. In a future of prolonged low flows, tour companies might find it harder to run such large boats safely, cutting off one main way to experience the canyon intimately.

Drought and low water aside, there's another aspect of the canyon's future that worries Victor R. Baker, a geologist at the University of Arizona. Dr. Baker has spent four decades exploring alcoves, high ledges and tributary mouths in the Colorado Basin. He scours them for the very particular patterns of sand and silt left by giant floods. The stories they tell are startling.

Mad cascades of water, ones at least as large as any the Grand Canyon experienced in the twentieth century, swept through it at least fifteen times in the past four and a half millenniums, Dr. Baker and his colleagues have found. Geological evidence upriver from the dam points to forty-four large floods of varying sizes there, most of them in the last 500 years.

As the atmosphere warms, allowing it to hold more moisture, the risk of another such deluge could be rising. If one struck when Lake Powell were already flush with melted snow, it could take out the dam, not to mention do considerable work on the canyon.

"I would think the future is going to be one moving toward, as they said in war, long periods of boredom interrupted by short episodes of total, absolute terror," Dr. Baker said.

None of the government agencies with a hand in managing the canyon can do much about that, not on their own. But they are trying to beat back some of the other forces remaking the canyon from within.

Since 1996, the Bureau of Reclamation, which owns Glen Canyon Dam, has occasionally released blasts of reservoir water to kick up sand from the riverbed and rebuild the canyon's beaches. The effects are noticeable. But the bureau conducts these "high-flow experiments" only when there's enough water in Powell to spare. In April, it held its first one in five years.

The National Park Service works to preserve the Grand Canyon's archaeological sites against erosion, even if that means leaving them swaddled in sand, where nobody sees them. "Those cultural resources that are covered by the sand are well suited by being covered by the sand," said Ed Keable, the park's superintendent.

Other issues, though, are so entrenched that addressing them just creates other problems. Take the spread of tamarisk, an invasive treelike shrub that has displaced native vegetation in the canyon and around other Western rivers. About two decades ago, officials decided to fight back by releasing beetles that loved eating tamarisk leaves. But the beetles loved those leaves so much, and their numbers grew so quickly, that they began threatening the Southwestern willow flycatcher, an endangered bird that nests in tamarisk.

There is a similar no-win feeling to the bigger question of how to keep the Colorado useful to everyone as it shrivels. The dam is the root cause of the canyon's environmental shifts, which also include big changes to fish populations. But simply allowing the river to flow more naturally through the existing dam, so water is stored primarily in Lake Mead instead of in both Mead and Powell, wouldn't reverse the shifts entirely.

Jack Schmidt, the director of the Center for Colorado River Studies at Utah State University, has concluded that the only way to allow sufficiently large amounts of sediment-rich water back into the canyon, short of dynamiting the dam, would be to drill new diversion tunnels into the sandstone around it. That would be costly, and require careful planning to dampen the immediate ecological effects.

"Like everything else in that damn river system," Dr. Schmidt said, "there's a consequence to everything."

It's the UC Davis scientists' sixth night on the Colorado, and it comes after several numbing hours of paddling against the wind. As the sun touches the canyon walls with the day's last glimmers of orange and gold, the graduate students sit in camp chairs chewing over what they've seen.

They are preparing for careers as academics and experts and policymakers, people who will shape how we live with the environmental fallout of past choices. Choices like damming rivers. Like building cities in floodplains. Like running economies on fossil fuels. Once, those were first-rate answers to society's needs. Now they require answers of their own—a whole wearying cascade of problems prompting solutions that create more problems.

"It becomes overwhelming," says Alma Wilcox, a master's student in environmental policy, sitting by a scraggly, haunted-looking grove of tamarisk. It helps, she says, to focus: "Having control over a really small aspect of it is empowering."

Yara Pasner, a doctoral student in hydrology, says she feels a duty to make sure the load on future generations is lessened, even if, or perhaps because, our forebears didn't do us that courtesy. "There's been a mentality that we will mess this up and the future generation will have more tools to fix this." Instead, she says, we've found that the consequences of many past decisions are harder to cope with than expected.

The next morning, the group floats into the realm of the canyon's oldest rocks. Almost two billion years ago, islands in the primordial sea crashed into the landmass that would become North America. The unimaginable heat and pressure from the collision cooked the rocks and sediment on the seafloor into layers of inky, shiny rock. This rock then lay buried beneath mountains that were formed in the collision, becoming squished and folded to create the otherworldly masses flanking the river today, which resemble nothing so much as freshly churned ice cream: dark gray schist swirled with salmon-pink granite.

But the mountains that sat above them? Those are all but gone, ground down over eons, their remnants long since scattered and recombined into new mountains, new formations.

"There were the Himalayas on top of this," says Nicholas

Pinter, the Davis geologist who has helped lead this expedition, gesturing from the end of a raft at Mile 78. "And it's eroded," he says. "Worn to an almost infinitesimally flat plane, before it all begins again."

Somewhere in among those grand happenings—within the tiniest, most insignificant-seeming snatches of geologic time—is the world we live in, the one we have.

Solving Climate Change Will Have Side Effects. Get Over It.

FROM *Los Angeles Times*

WHEN I WROTE a column two weeks ago urging the Biden administration to approve a lot more solar and wind farms on Western public lands, I knew I would get flak from critics of large-scale renewable energy—and indeed I did.

On social media, conservationists blasted me for what they described as my failure to understand that sprawling solar projects and towering wind turbines tear up wildlife habitat and destroy treasured landscapes. They called me a shill for money-grubbing utility companies and suggested it's obvious that we should rebuild our energy systems around solar panels on rooftops.

I'm sympathetic to those arguments and want to clarify where I'm coming from.

I'm familiar with the science showing that human survival depends in part on limiting further biodiversity loss and protecting much of the remaining natural world. I feel a deep appreciation for America's spectacular public lands; I've hiked and camped across the West, from the Teton Crest Trail to Mt. San Jacinto. I'd love to see more national monuments created.

In an ideal universe, I'd support building renewable energy exclusively within cities and on previously disturbed lands such as farm fields and irrigation canals. In an ideal universe, I'd support only climate solutions that don't cause other problems.

But we don't live in an ideal universe.

We live in a universe where every clean energy technology has drawbacks, whether economic or technical or political. A universe where there aren't enough rooftops to replace all the fossil fuels we now burn. Where skeptical farmers are fighting to stop their neighbors from switching to solar energy production. Where building solar on canals is wildly expensive, at least so far.

Just as importantly, we live in a universe where human beings use mind-boggling amounts of energy.

Every time we flip a light switch, run the dishwater or take a drive, we're using energy. Our coffee mugs, our clothes, our homes—they took energy to manufacture. Same with the food we eat, the TV shows we love and our favorite board games.

Even with aggressive energy-efficiency improvements, we'll need an unprecedented solar and wind building spree to replace all the coal, oil and fossil natural gas boiling the planet and spewing toxic fumes responsible for millions of deaths each year.

So why have we had so much trouble coalescing around the need for a broad range of clean energy technologies?

If you ask me, it's because it's so hard to grapple with the enormity of the climate crisis.

Deadlier heat waves, bigger wildfires, shrinking reservoirs, rising oceans—we understand them on paper. But most of the time they're abstract, lurking in the background. Whereas a wind farm that will kill golden eagles is tangible, easy to grasp. Same with a solar farm that will be visible from Joshua Tree National Park, or an electric line that will cut through ancient burial sites.

It's not wrong to care about that stuff. It's not wrong to want to protect the places we know and love.

But too many of us have gotten stuck looking at the world through a narrow defining lens.

Mistrustful of monopoly utility companies? Then you probably see rooftop solar panels as the ideal climate change solution. Live near the coast and love the ocean views? Then solar farms in the desert probably sound better than offshore wind turbines. Find it easier to cope with the idea of climate chaos if you can convince yourself a single technology or policy will fix everything? Then maybe you're a devotee of nuclear reactors, or a carbon fee, or carbon capture and storage.

If we were having this conversation a few decades ago—say in

1988, after climate scientist James Hansen testified to Congress that global warming had arrived—then debating the best suite of climate solutions might be a good use of time. We could work together to reach consensus on the right path forward and ensure the side effects were as painless as possible.

But this is 2023, not 1988.

Largely thanks to the fossil fuel industry's climate denial and the Republican Party's continued intransigence, we're out of time. I keep saying this in my columns, but it bears repeating: Scientists have calculated we need to cut global climate pollution nearly in half by 2030, just seven years from now, to avoid an extremely scary future. Seven years is nothing. This is an emergency.

Much as I hate the idea of paving over desert tortoise habitat with solar panels or refusing to remove dams that have decimated salmon populations, I hate the idea of 3 degrees Celsius of planetary warming a lot more. Much as I sympathize with rural towns that don't want to live with industrial wind turbines as their neighbors, I sympathize more with my neighbors here in Los Angeles who can't afford air conditioning and don't want to die of heatstroke the next time the thermostat hits 121 degrees.

For those of you reading this with frustration—I realize I'm probably not going to convince you. You don't know why I can't just understand that your climate solution is the best one, and use my platform as a journalist to help bring it about.

My unsatisfying response is that I'm a realist.

I know that not every proposed clean energy solution is a good idea. But the reality is that solar farms and wind turbines, for all their faults, are some of the most proven, cost-effective, politically popular tools for reducing our reliance on fossil fuels.

I could spend all my time singing the praises of rooftop solar—which I did last week, by the way—and it wouldn't change the fact that avoiding the worst consequences of climate change will be a hell of a lot easier if we embrace big solar and wind.

Now, for those of you reading this and nodding in agreement—thanks for your support. But I hope you'll stop and ask yourself: What are you personally willing to sacrifice to bring about a safe climate future? What changes will you make in your life?

Will you eat less meat, replace your gas stove with an induction cooktop or lease an electric car? Will you make climate change a top priority at the ballot box, and post about it on Instagram, and bring it up at the dinner table on Thanksgiving?

If you hear about a climate solution that rubs you the wrong way, will you swallow hard and look the other way?

Because that's what it's going to take.

To maintain a habitable planet for ourselves and our children and grandchildren, we'll need to make some compromises. We'll need to stand by and watch as some pristine ecosystems are razed in the name of renewable energy. We'll need to learn to live with exorbitantly wealthy investors raking in additional profits at our expense. We'll need to elect some politicians whose ideas don't fully line up with our own, because they're nonetheless our best hope of avoiding planetary collapse.

Above all, we'll need to stop yelling at each other and start co-operating with people we think are wrong.

That's the world we live in. Welcome to the Anthropocene.

Buried Under the Ice

FROM *Washington Post*

GREENLAND HAD NOT been kind to Joerg Schaefer. For twenty-one days, he had endured howling winds and blistering cold. He sampled ice until his fingers went numb and shoveled snow until his shoulders burned. His shelter was a cramped yellow tent, his bathroom a hole in the ground.

And the experiment that brought him to this ruthless, frozen expanse—an unprecedented effort to drill through more than 1,600 feet of ice and uncover the bedrock below—was teetering on the brink of failure.

A thin, dark fracture had appeared deep in the ice, making it impossible to drill. The team's engineers were trying to repair the hole, but it was far from clear whether the fix would work. Schaefer's hopes—and research that could help predict the fate of Greenland's ice sheet and its contribution to rising seas—clung to the smallest possible chance that the team might still find what they came for.

Now, Schaefer would have to leave with his mission unfinished. Family obligations had called him home to New York, and a helicopter was waiting to take him back to civilization.

"The timing feels completely wrong," said Schaefer, a climate geochemist at the Lamont-Doherty Earth Observatory and the lead investigator for the project known as GreenDrill. "We have been working toward getting that bedrock for literally years. . . . It's pretty heartbreaking now to leave."

The drilling expedition last spring was only the third time in history that researchers had tried to extract rock from deep beneath Greenland's ice. Scientists have less material from under the ice sheet than they do from the surface of the moon. But Schaefer believes the uncharted landscape is key to understanding Greenland's past—and to forecasting the future of the warming Earth.

The Greenland Ice Sheet contributes more to rising oceans than any other ice mass on the planet. If it all disappeared, it would raise global sea levels by 24 feet, devastating coastlines that are home to about half the world's population. Yet computer simulations and modern observations alone can't precisely predict how Greenland might melt. Researchers are still unsure whether rising temperatures have already pushed the ice sheet into irreversible decline.

Greenland's bedrock holds the ground truth, Schaefer said. The ancient stone that underlies the island is solid, persistent and almost unmovable. It was present the last time the ice sheet completely disappeared, and it still contains chemical signatures of how that melt unfolded. It can help scientists figure out how drastically Greenland may change in the face of today's rising temperatures. And that, in turn, can help the world prepare for the sea level rise that will follow.

Schaefer trudged to the drill tent to say goodbye. Inside, four engineers were troubleshooting yet another problem with their massive machine.

"So, you're leaving us?" asked Tanner Kuhl, the project's lead driller.

"I'm not thrilled about it," Schaefer said. He clapped Kuhl on the shoulder. "Keep fighting."

The team was running out of time and money to get through the last 390 feet of ice. Success would require drilling faster than they ever had before. But no one was prepared to give up—not yet. Too much had been invested in their five-year, multimillion-dollar project to collect bedrock samples from around Greenland. Too much was at stake, as human-made pollution warms the Arctic at a pace never seen before.

Reluctantly, Schaefer clambered into the helicopter. He watched as the ground fell away, and the GreenDrill encampment shrank to a tiny splotch of color amid the endless white.

He could only imagine what secrets lay beneath that frozen

surface. He could only hope—for his own sake, and for the planet's—that those secrets would someday be known.

The dome

The morning of Schaefer's arrival on the ice sheet, three weeks earlier, was luminous and still. Barely a breeze jostled the helicopter that carried him to the field camp in northwest Greenland. Beneath him, the glittering expanse of the Prudhoe Dome ice cap resembled the ocean: Beautiful. Mysterious. Mind-bendingly huge.

Theoretically, Schaefer had already grasped this. The dome's size was one reason it had been selected as a GreenDrill sampling site—the first of three strategic locations around Greenland where he and his colleagues planned to drill for bedrock samples over the next few years. Yet Schaefer was still stunned as the helicopter touched down on the dome's summit. Despite a decades-long career in polar science, this was his first time on an ice sheet.

"It's a monster," he thought. Schaefer tried to imagine all that ice melting and flowing into the ocean. The catastrophic possibility made him shiver.

He knew that such a disaster had happened before. In 2016, Schaefer and his close colleague Jason Briner, GreenDrill's co-director and a geology professor at the University at Buffalo, were part of a team that analyzed the single bedrock sample that had been previously collected from beneath the thickest part of the ice sheet. The rock contained chemical signatures showing it had been exposed to the sky in the past 1.1 million years. In a paper they published in the journal *Nature*, the scientists concluded that almost all of Greenland—including regions now covered by ice more than a mile deep—must have melted at least once within that time frame.

"That was a game changer," said Neil Glasser, a glaciologist at Aberystwyth University who has followed Schaefer's research. "It said that the Greenland Ice Sheet is far more dynamic than we had ever thought."

The findings ran counter to many scientists' belief that Greenland has been relatively stable throughout recent geologic history,

as the Earth oscillated between ice ages and milder warm periods
known as interglacials. If the ice sheet could melt at a time when
global temperatures never got much higher than they are now, it
was a worrying harbinger of what ongoing human-caused warming
might bring.

For Schaefer and Briner, the discovery also underscored how
bedrock could complement findings from ice cores—the long
slices of frozen material that had traditionally been the focus
of polar science. For decades, scientists had studied air bubbles
trapped in ice to uncover crucial clues about the climate at the
time the ice formed.

Yet ice cores, by their very nature, could only reveal what happened during the colder phases of Earth's history. They couldn't
answer what Schaefer said is arguably the most important question
facing humanity now: "What happened when it got warm?"

With GreenDrill—which is funded by the U.S. National Science
Foundation—he and Briner aimed to open up a new kind of record, one that didn't melt away. By collecting bedrock samples
from around the island, they could gain a clearer picture of exactly when the ice sheet last vanished and what parts of Greenland
melted first.

But to get those rocks, they would need to survive for more
than a month in one of the most remote and hostile places on
Earth.

The effort was shaping up to be harder than anyone had antic-
ipated. It took seven plane trips for the team to haul GreenDrill's
equipment onto the ice sheet. Twice they were delayed by bliz-
zards. Then a windstorm halted work on the drill for three full
days.

The scientists who oversaw camp setup—Lamont Doherty geologist Nicolás Young and University at Buffalo Ph.D. student Caleb
Walcott—already looked exhausted as they welcomed Schaefer to
the ice.

They watched Schaefer's helicopter depart in a whirl of wind-
blown snow. Then Young and Walcott led him to one of about a
dozen yellow tents that had been staked onto the ice—Schaefer's
home over the next three weeks. Its insulated double walls were
designed to keep temperatures inside just above freezing. A cot
lined with an inflatable sleeping pad would keep him off the frigid
tent floor.

Next door stood the larger kitchen tent, which was equipped with propane stoves and a single space heater—the only heat source they would have on the ice. The nearby storage tent was stocked with hundreds of pounds worth of beans, granola bars and other nonperishable foods.

Dozens of green, black and red flags marked paths through the camp; in case a blizzard made it hard to see their tents, the team members were to follow the flags to safety.

As they headed toward the drill tent, Schaefer heard the growl of engines and the whine of machinery. Yet he was unprepared for what he saw when he peered inside.

The Agile Sub-Ice Geological (ASIG) drill was a complex beast: 30,000 pounds of cable, aluminum and steel.

The machine was state of the art; whereas older drills would take years to cut through 1,600 feet of ice, ASIG could achieve the same feat in a matter of weeks. It was one of few drills in the world that could extract bedrock from deep beneath an ice sheet. Yet it had been used successfully just once before, and never in Greenland.

For the first time—but not the last—Schaefer found himself wondering, "What on Earth did we do to get this enterprise up here?"

"Maybe something goes wrong?" he thought. "Maybe we don't get through this ice."

But he was reassured by the unflagging support of the National Science Foundation and the unflappable competence of the GreenDrill team. Briner, the project's co-director, was a leading scholar of Arctic ice. Kuhl, an engineer with the U.S. Ice Drilling Program, had overseen the only other successful ASIG project, in Antarctica. His partner, Richard Erickson, had more than two decades of experience in minerals drilling. The camp manager, technician Troy Nave, had spent a combined thirty-four months living at the poles.

And then there was Allie Balter-Kennedy, Schaefer's former Ph.D. student who had defended her dissertation just a month earlier. Though she was many years his junior, Balter-Kennedy had spent much more time doing remote field work. It was she who explained to Schaefer how to stay warm at night—by sticking a bottle of boiling water in his sleeping bag—and who tempered his frustration every time a storm threw a wrench in their plans.

"They are just the ultimate professionals," Schaefer often said of his colleagues. If any group could pull off such a far-fetched experiment, he thought, this one would.

The quest

When Schaefer thought about the research that had led him to this remote corner of the planet, it struck him as something out of a fairy-tale.

The story began a long time ago, in a distant corner of the universe, amid the death throes of ancient suns.

When stars explode, they send sprays of high energy particles into the cosmos. A few of these cosmic rays are able to make it through Earth's atmosphere and reach the ground below. And when those particles encounter rocks, they can interact with certain elements to create rare chemicals called cosmogenic nuclides.

These nuclides accumulate in surface rocks at predictable rates. Some are also radioactive, and they decay into new forms on distinctive timelines. This allows scientists to use them as molecular clocks. By counting the numbers of nuclides within a rock, scientists can tell how long it has been exposed to cosmic ray bombardment. And by comparing the ratios of various decaying elements, they can determine when ice began to block the rock's view of the sky.

With this technique, ordinary rocks are transformed into something almost fantastic. They are witnesses, capable of remembering history that occurred long before any humans were around to record it. They are messengers, carrying warnings of how the ice—and the Earth—could yet transform.

"It's like a magic lamp," Schaefer said. "It's pretty crazy that a little piece of rock can tell you the story of this massive sheet of ice."

But cosmogenic nuclides are difficult to study; Schaefer had spent much of his career helping to develop the equipment and techniques needed to detect a few tiny nuclides in a piece of ancient rock.

And—as Schaefer was now learning—rocks from beneath an ice sheet are immensely challenging to obtain.

Days at the GreenDrill field camp began at 6:30 a.m., as Schaefer woke to the sound of engineers turning on the ASIG generators.

He dressed quickly in the cold, and grabbed coffee and breakfast in the kitchen tent. With a satellite phone, he checked in with the National Science Foundation support team and got an updated weather forecast: Usually windy. Sometimes snowy. Almost always well below zero degrees Fahrenheit.

Then he bundled back up and headed outside to start work.

Until the drillers hit bedrock, the scientists' primary task was to collect the ice chips pushed up by the Isopar drilling fluid. Schaefer, Young and Balter-Kennedy took turns scooping the ice into small plastic bottles. Back at their lab, they would analyze those samples for clues about Greenland's temperature history— research that would complement their bedrock analysis.

The rest of their time was spent shoveling in a Sisyphean effort to keep the camp from being buried by the endlessly swirling windblown snow.

Fifteen miles away, Briner and his student Walcott had set up another camp closer to the edge of the ice sheet. There, the team was using a smaller Winkie drill to collect a second sample from beneath a much thinner part of the ice cap.

In each sample, the team planned to analyze as many as six different cosmogenic nuclides, but they would focus on two: beryllium-10 and aluminium-26.

When cosmic rays strike a rock, radioactive aluminum accumulates at a much faster rate than beryllium. Yet aluminum-26 also decays faster once the rock has been covered up by ice. If a sample has relatively little aluminum-26 compared to beryllium-10, it suggests that site has been buried under ice for hundreds of thousands of years. But if Schaefer and Briner found a high ratio of aluminum-26 to beryllium-10, it would mean the site had been ice-free in the more recent past.

With that analysis, the researchers would then be able to compare the exposure histories of each drilling site. They could determine how much warming it previously took for the ice sheet to retreat the 15 miles between Winkie and ASIG, which would give them a sense of how fast Prudhoe Dome might disappear someday.

"That's a really important element of GreenDrill," Briner said. "We're drilling more than one site so we can look at different parts of the ice sheet . . . and we can ask, 'What does it say about the shape of the whole ice sheet if this part is gone?'"

As Schaefer settled into the rhythm of camp, his body adjusted to the incessant polar sunlight and the relentless manual labor, leaving his mind free to wander. Often, he found himself fixating on the ice sheet—its incomprehensible vastness, its unfathomable fragility. How could something this immense and forbidding be so vulnerable? How could a landscape so obviously capable of killing a person be at risk because of humanity?

Yet in the years since Schaefer and Briner published their 2016 paper, Greenland's vulnerability had become more and more clear. The melt season of 2023 was shaping up to be among the island's worst, with the ice sheet on track to lose roughly 150 billion tons of ice by summer's end.

Meanwhile, research showed how the melting process is self-reinforcing: Dark pools of water on Greenland's surface absorbed the sun's heat, rather than reflecting it. The diminishing height of the ice sheet exposes the surface to the warmer air at lower altitudes. If the ice shrinks far enough, it could enable the rising ocean to infiltrate the center of the island, which is below sea level. That warmer water would melt the ice sheet from below, accelerating its decline.

Under the worst case warming scenarios, the melting Greenland ice sheet is expected to contribute as much as half a foot to global sea level rise by the end of the century. It would swamp Miami, submerge Lagos and deluge Mumbai. In New York City, where Schaefer lives with his wife and two children, coastal flooding would inundate the subways, destroy sewers and power plants and wash away people's homes.

This knowledge made Schaefer's quest feel less like a fairy-tale than a medical drama—an urgent effort to understand the sickest part of Earth's embattled climate system.

"We are basically taking biopsies," he said, "which will hopefully tell us how sensitive the patient is to ongoing warming—and how much warming is fatal."

The fracture

As soon as he saw the looks on the drillers' faces, Schaefer realized something had gone terribly wrong.

For three weeks, they'd been making slow but steady progress through the ice sheet, reaching nearly 1,300 feet below the surface.

But suddenly, the pressure in the borehole started to drop. The Isopar wasn't able to push the ice chips up through the drill rod, and Kuhl feared the fluid must be leaking through a crack in the hole.

Schaefer knew the word "fracture" was like "Voldemort"—the name of the villain in the Harry Potter series. It inspired so much dread that the scientists avoided even mentioning it.

Yet there was no denying the thin, dark fissure in the ice that appeared about 250 feet down when Schaefer dropped a camera into the borehole. His heart sank. If they couldn't find a way to isolate the fracture, they wouldn't be able to maintain enough pressure to push chips out of the hole, and drilling would grind to a halt.

"Basically game over," Schaefer wrote in his waterproof field notebook.

He went to bed that night with his mind racing. The fracture shouldn't have happened, he thought. Ice was supposed to be able to withstand pressures far greater than what Kuhl and Erickson were using. Did this mean the ASIG drill didn't work as well in Greenland? Would it threaten all of the other drilling they had planned?

The next morning, he woke to the howling of the wind.

Another unseasonable blizzard had descended on Prudhoe Dome, blotting out the endless icescape with an impenetrable cloud of white. When Schaefer peered out of his tent, he could barely see the colored safety flags that were supposed to guide him through camp.

The team gathered in the kitchen tent, listening to the eerie whistle of snow blowing over ice and the snapping of flags in 50-miles-per-hour gusts.

Schaefer got a satellite message from Briner: Conditions were even worse at the Winkie Camp, where the fierce wind had ripped a hole in their kitchen tent, scattering supplies and burying cooking equipment in snow.

It wasn't safe to work on the fracture. It wasn't safe to even be outside for more than a few seconds. All they could do was wait for the storm to end.

Schaefer tried to combat the growing worry that he had led his team into disaster. He had underestimated the complexity of drilling through so much ice. He hadn't reckoned with the possibility of so much bad weather.

In his lowest moments, it felt as though Greenland had been conspiring against him.

But after three days, the wind finally died down. The air cleared. And Kuhl had come up with a plan to fix the fracture.

The engineers could deepen the upper part of the borehole, which was sealed with the aluminum casing, so that the casing extended to cover up the crack. They could also switch the way Isopar circulated through the borehole to minimize pressure on the ice.

It took the drill team two days to make the repair. A week after the fracture first appeared, they tried refilling the borehole with fluid. It held. The project was still alive—but just barely.

Schaefer got more good news the following day, when Briner arrived from the Winkie camp.

"I brought you a present," Briner said, grinning. He opened a long cardboard box to reveal 79 inches of sediment and pebbles mixed with chunks of a large bolder. Despite the blizzard, the Winkie team was able to extract their rock core from beneath 300 feet of ice.

"Are you kidding me?" Schaefer gasped. The sediments could include traces of plants and other clues to the ancient Greenlandic environment. And the boulder pieces contained plenty of quartz—perfect for cosmogenic nuclide analysis. It would give the team crucial clues about what had happened at the ice sheet's vulnerable outermost edge.

Schaefer reached out to shake hands with Elliot Moravec, the lead Winkie driller. "Thank you, man," he said. "That's a really great sample."

With the Winkie core in hand, Moravec and his partner Forest Rubin Harmon were also free to help Kuhl and Erickson keep the ASIG machinery running.

Yet now the team had another enemy: time. There was less than a week until they were supposed to start taking down the camp. They also had to redrill through the frozen ice shavings that fell into the borehole during the repair process. Yet they couldn't go too fast, or they risked triggering another fracture.

"It's slow motion, drilling through the slush," Kuhl said, shouting over the whine of the drill.

Then the noise stopped. Kuhl and Erickson peered into the borehole: They'd hit another blockage.

Kuhl grimaced. "There's almost no chance we get this done now," he said. "But who knows? We'll keep trying and maybe something will work."

By the time Schaefer was set to depart, progress was still painfully slow.

"Leaving these guys alone now feels really shitty," the scientist said. But his family needed him at home, and flights out of Greenland were so infrequent there was no option for him to take a later plane.

Schaefer shook his head. Three weeks prior, he thought this was a medium-risk, high-reward project—that the effort and expense were absolutely justified by the scientific value of the rocks they would uncover.

Now, knowing the true level of risk, facing down the very real probability of failure, he couldn't avoid the question: Was it all worth it?

The rocks

The Manhattan streets were packed with people, ringing with conversation and birdsong. The air was warm. The late spring breeze was gentle.

Yet Schaefer felt ill at ease as he paced the rooms of his narrow New York apartment. His mind was still 2,500 miles away.

Back at Prudhoe Dome, Balter-Kennedy was overseeing the last-ditch efforts to salvage GreenDrill's field season. She sent Schaefer daily updates via satellite messenger, but the news wasn't good. On that Friday, the drillers only managed to cut through about 40 feet of ice. On that Saturday, they pulled up the entire apparatus and switched to a different type of drill rod—to no avail.

That Sunday morning in New York dawned sunny and mild. In Greenland, it was the last day before planes were supposed to start taking their equipment off the ice. The complex operation could not be delayed—and even if that was an option, conditions would soon become too warm for the team to safely drill.

By the time his phone buzzed with Balter-Kennedy's regular evening check-in, Schaefer was braced for failure. He had already rehearsed what he would tell the team: This didn't change how hard they worked, or how much he appreciated them. They'd tried to do something no one had done before. They had learned valuable lessons for 2024. And the world desperately needed the information buried beneath Greenland's ice—no matter the risk, their efforts were worth it.

But when Schaefer answered the satellite call, Balter-Kennedy was ecstatic. The team got through more than 150 feet of ice that day—an all-time record for the drill they were using. They were 90 percent of the way to bedrock.

Schaefer felt as though his heart had stopped beating. The rest of the conversation was a blur, as Balter-Kennedy explained how the engineers finally overcame the blockages that had slowed their progress. Erickson, the longtime minerals driller, suggested they try using a drill bit that is usually reserved for rocks. It was an outside-the-box suggestion from someone unaccustomed to drilling through ice—but it worked.

"The ASIG is back in the race," Schaefer said that night. "And with all the overwhelm and confusion that I have inside me, it's clearly much, much better than the big emptiness I would have if it would be called off."

Still cautious, he added: "Which of course, can still occur. The slightest problem now and it's over. But they are within striking distance. It could still happen."

The next day, Schaefer had to hold himself back from bombarding Balter-Kennedy with requests for updates. Instead, his thoughts drifted to the laboratories at Lamont Doherty where his team would analyze the GreenDrill bedrock—if they ever got it.

He could already imagine the steps that would come next.

First, the scientists would slice samples from the main core. Next would be a weeks-long chemical procedure designed to rid the sample of unwanted material, leaving behind only the elements the scientists had chosen to analyze. Finally, the samples would be sent to labs in California and Indiana, where they'd be shot through particle accelerators and weighed to determine the proportion of cosmogenic nuclides.

Within about six months, the GreenDrill team would know when the ASIG bedrock last saw daylight. And those results would illuminate whether Prudhoe Dome's future is simply grim—or completely catastrophic.

Schaefer hoped he would find that the ASIG site hadn't been ice-free since the interglacial periods that punctuated the Pleistocene epoch, more than 100,000 years ago. But it was possible that this region melted during the Holocene, the 11,700-year stretch of mild temperatures that began at the end of the last ice age and continues through today.

"That would be another level of devastating," Schaefer said. Modern temperatures are quickly surpassing anything seen during the Holocene epoch. If Prudhoe Dome couldn't survive those conditions, then under human-caused climate change, it may soon be doomed.

Five days after Schaefer's departure, ASIG hit the bottom of the ice sheet—1,671 feet below the surface. Now the drillers would switch to a bit that could cut cores out of rock, while Balter-Kennedy and Walcott scrambled to pack up the rest of the camp. All Schaefer could do was wait.

He alternated between agitation and exhaustion—first pacing his apartment, then collapsing on the couch, then springing up to pace again. On a whim, he wandered into the soaring Gothic sanctuary of the Cathedral of St. John the Divine. It had been Balter-Kennedy's idea: when he complained of his helplessness on one of their satellite calls, she jokingly suggested he could pray for the team.

So, although it had been years since he set foot in a church, he paused to light a candle for GreenDrill.

Finally, in the very early hours of a Wednesday morning, he got the messages he had been waiting for.

"We have 3 m of bedrock!!!"

"Going down for 4.5 now"

Hands shaking with adrenaline, Schaefer could barely type his response.

"Ukidin?? Not real! Omg! YESYEYES!! WowFantastic!!!U literally change Ice/Sheet Science! All undercyour lead! That's why u joined!"

"Gogogo!"

A few hours later, Balter-Kennedy called on the satellite phone. In her usual, even-keeled tone, she described how the drillers worked through dinner and deep into the night. They had nearly 10 feet of sediment and another 14.5 feet of pristine bedrock—more material than Schaefer even dreamed was possible.

But when she hurtled into questions about departure logistics, Schaefer interrupted her. He wanted to make sure she had a chance to appreciate the moment. "You guys just opened a new era," he said.

Afterward, he pulled up the voice notes app on his phone.

"What can I say?" he murmured into the microphone. "Science—it's not well paid. It has a lot of problems. But it's so fulfilling."

Schaefer thought of the bedrock that Balter-Kennedy described to him: gorgeous cylinders of pale gray quartz and glittering chunks of garnet. These were their "moon rocks," he said—the first material from an unexplored landscape.

It seemed fitting that the project's youngest scientists—Balter-Kennedy and Walcott—oversaw the breakthrough. They had poured so much of themselves into this experiment. Their careers, and their lives, would be shaped by what was learned.

And now that GreenDrill had proven it was possible, the next generation of researchers would be able to collect even more samples from underneath the ice. They could peer into history no human had witnessed. They could start asking the questions that only rocks can answer.

What they find will almost certainly be frightening, Schaefer knew. It will shed new light on the dangers facing humanity. It will highlight the urgency of tackling problems science alone can't solve: how to transition away from polluting fossil fuels, how to protect vulnerable people from climate disasters like sea level rise. There was far more work to be done.

But for now, in the darkness of his apartment, he simply basked in the joy of discovery.

"It all worked out," he said, his voice slurred by tears and exhaustion. "And the entire, the entire—" he paused. "It was worth it."

DOUGLAS FOX

Rogues of the Rainforest

FROM *bioGraphic*

BORIS BERNAL VARGAS often begins work with a forty-minute commute. Starting from his home in the village of Las Pavas, Panama, he walks through a pasture to a wooden dock on Gatun Lake. From there, he and a dozen other workers skim across a couple kilometers of water in a motorboat to Barro Colorado Island—a refuge of tropical forest that in some places has not been logged for hundreds of years. Beneath the canopy, workers follow a short path up a gentle slope, and at last, wade through ankle-deep leaves to a row of PVC pipes driven into the ground.

The forest is chaotic and tangled, seemingly impervious to the human impulse to catalogue and name. But past this row of markers, every tree larger than a person's pinkie is tagged with a six- or seven-digit number. This 50-hectare swath contains roughly 350,000 trees—about three times the number of hairs on a person's head. And for forty years, Bernal Vargas and other researchers have measured the growth of each for the Smithsonian Tropical Research Institute (STRI), based in nearby Panama City.

Trees aren't the only creatures of interest here. Starting in 2016, Bernal Vargas also investigated a long overlooked kind of plant: lianas, vines with thick, woody stems that climb tropical trees—often spiraling up their trunks—in search of sunlight. Within this plot, every liana above the same minimum pinkie thickness is also tagged. For twenty-three months, Bernal Vargas spent his days identifying each, gauging the diameter of its stem with calipers or tape measure, and penciling the results on a clipboard. The vines were so tangled that Bernal Vargas could often measure only a few dozen per day—an agonizingly precise, Sisyphean task, punctuated by moments of beauty and irritation: the sight of a golden puma passing silently; the searing stings of inch-long bullet ants that use the vines as highways; the droplets of urine spread by agitated howler monkeys whooping in the branches above.

But with persistence, Bernal Vargas and his coworkers eventually measured every liana on the site. It was the second time they had done this, and it confirmed a mysterious and troubling trend. During the past decade, the number of lianas had increased by 29 percent on Barro Colorado Island, reaching a record 117,100 individuals. Researchers have seen similar increases in dozens of other study plots scattered across the tropical forests of Central and South America.

Lianas seem like footnotes to the forest—scrawny weaklings dominated by the stout-trunked trees all around them. "When I started in the mid-90s, nobody cared about lianas. They're just these things you trip over and cut with a machete," says Stefan Schnitzer, an ecologist who employs Bernal Vargas and the other workers measuring these vines.

But viewed from above the canopy, lianas are the ones that dominate. A single vine can spread across dozens of trees, unfurling its leaves like parasols over those that grow more slowly. In the race to grab precious sunlight, lianas are literally outrunning and outmaneuvering trees, says José Medina-Vega, a former postdoc of Schnitzer's who now works at the Smithsonian Global Earth Observatory Network in Washington, D.C. "They have a guerrilla strategy."

Scientists consider tropical forests critical allies in the struggle against climate change because they have the capacity to absorb and store billions of tons of anthropogenic CO_2. But rising CO_2 and temperatures are already causing rainforest trees to die younger, and lianas may compound the problem. They "have this really strong negative effect," says Schnitzer, who has joint appointments at STRI in Panama and Marquette University in Milwaukee. "They prevent trees from taking up as much carbon." They reduce tree growth. And they can send trees crashing prematurely to the ground.

Scientists have long thought that anthropogenic environmental changes trigger liana expansion: Perhaps rising levels of CO_2 or atmospheric deposition of nitrogen and phosphorus have spurred the vines' growth. But the exact mechanism has remained mysterious for twenty years. Some researchers even wonder whether it may be the misleading result of biases in the way vines are studied. Mounting evidence, though, suggests that the spread of lianas is not only a symptom of the ways that climate change is remaking tropical forests, but also, increasingly, a cause.

The competition among trees and lianas is not a battle between unrelated plants. Oddly, it grew out of an evolutionary struggle between tree species.

Coiling, woody vines, many lianas are actually descended from trees that forsook the Puritanical labor of building sturdy trunks and instead opted to slink up the trunks of other trees. This transformation to "structural parasite" has occurred many times, across a quarter of all plant families, sometimes pitting members of the same family against each other. And it truly is a transformation, because it involves much more than a slimmer trunk.

Because a liana's stem does not need to support weight, the structure of its wood can maximize a different function: the transport of water from its roots to its leaves. Cutting through a liana stem often reveals a cross section ornately dotted, like the inside of a kiwi fruit. Those dots are water-sucking tubes, called xylem. Because they are wider than those in trees, they produce less friction with the water rising through them, vastly increasing the volume that can flow and allowing much more rapid photosynthesis and growth.

For over a century, researchers largely ignored lianas. But in the past few decades, they have finally begun to realize that lianas are important in their own right—powerful plants that can shred the social fabric of forests. It started in the late 1970s when a young graduate student named Francis Putz arrived for a fellowship at Barro Colorado Island, intent on studying how these resource thieves slither through the canopy, groping their way sightlessly from tree to tree. Using a modified .22 rifle, he would shoot fishing line over a branch 20 to 40 meters above, use that line to pull up a nylon cord, and use the cord to pull up a climbing rope. Donning a harness, he spent hundreds of hours dangling in the canopy—taking occasional afternoon naps—and familiarizing himself with the slow, bare-knuckle battles between climbing lianas and trees that did not want to be climbed.

Putz discovered that lianas are contagious. Once one managed to climb a tree, it sent shoots to grasp neighboring trees—not unlike Putz's fishing lines. This contagion could spread surprisingly far. One liana, an entada pea vine with spiraling, meter-long pods, had spread across the tops of forty-nine trees.

Putz also saw how vigorously trees fought to avoid liana infestation. Some defended themselves with leaves and branches that easily peeled away, sending grasping vines crashing to the ground. Palms developed serrated sword leaves that could slice through young vines. Other trees had more violent means. In one frightening episode, Putz was 20 or 30 meters up a stout-trunked tabebuia tree when he saw a squall gathering in the distance. Having read John Muir's account of riding out a storm high in a sequoia, he figured it was worth a try. It was, he says, "the kind of thing you do when you're in your twenties."

The winds arrived with the roar of an oncoming train and the screams of agitated monkeys. And for about ten minutes, Putz jolted and swung on his ropes as the neighboring tree—a slender zanthoxylum with a thorny trunk and fern-like leaves, connected to his own by three different lianas—whiplashed, yanking his tree this way and that. When the wind subsided, he realized he had wetted himself. He also noticed that all three vines had broken. Putz came to see that the tree's pliancy was defensive. In addition to breaking lianas, its mosh-pit thrashing stripped branches from neighboring trees—creating a buffer zone that the vines would find difficult to cross.

The stakes were clear: When lianas did manage to tether trees together, one falling individual could pull down the rest, killing them all. The lianas, though, could capitalize on the newly intense sunlight to grow rapidly over the clearing. These rapidly growing vines thrive wherever the sunlight is intense—whether at the top of the canopy, or in a clearing on the ground.

At the time Putz conducted this research, scientists generally assumed that these ecological battles were in equilibrium on a broad scale in forests that had little direct contact with industrial society. If lianas overran one spot, trees often balanced it out by dominating elsewhere. But another scientist's work soon began to undercut this idea.

In the early 1990s, a Ph.D. student at the Missouri Botanical Garden named Oliver Phillips was attempting to answer a long-standing question in ecology: Did forests with faster turnover—that is, faster growth, death, and replacement of trees—also have higher levels of biodiversity? To find out, Phillips had assembled records of tree growth and death from forty tropical forest plots in Asia, Africa, Australia, and North and South America. Most had been surveyed only twice over ten to fifteen years, but collectively, they spanned from 1934 to 1993. And as Phillips tabulated the results, something strange emerged. The rate of tree turnover seemed to be increasing.

Up until 1960, about 1 percent of the trees in any given plot died and were replaced by saplings each year. But by the late 1980s, the rate had doubled to almost 2 percent. "I wasn't looking for a pattern," recalls Phillips. "The data were talking"—and they seemed to be describing a global change. Perhaps rising CO_2, which provides the building blocks for trees to form carbohydrates during photosynthesis, was pushing trees to grow more quickly and die young. Or perhaps climate change's increasing temperatures and droughts were killing trees early. Or stronger windstorms were simply felling more trees.

Whatever the cause, Phillips also noticed something else. Six of the plots, located in Peru, had tracked the growth of lianas since 1983, and in five, lianas seemed to be increasing. "It's quite clear how lianas can and do accelerate tree mortality," says Phillips, who is now at the University of Leeds in the United Kingdom. When he looked at dozens of other sites across the Amazon, he found a widespread trend: Lianas were growing more abundant at the roaring rate of 1 to 5 percent per year. Over twenty years, they had nearly doubled.

That 2002 discovery made a big splash. "No one believed it at first," says Schnitzer, who leads the liana research team at the STRI in Panama. Some scientists wondered whether the researchers were simply getting better at counting lianas. But other studies found similar results.

Around that time, Schnitzer himself was drifting toward lianas. He had just finished his Ph.D. dissertation at the University of Pittsburgh, showing that falling trees create gaps in tropical forests that boost biological diversity, by providing space where fast-growing "pioneer" tree species can root. Eventually, through a process called succession, those pioneers give way to slower-growing species that make up the bulk of the forest. But Schnitzer noticed (just as Putz had before him) that lianas also often thrive in these sunny gaps; and what's more, he found that lianas can actually overwhelm the pioneer trees, shading them and slowing their growth. Gaps that would normally fill within five to ten years often remained sparsely treed for twenty-five years or more. The lianas, he says, are "holding disturbed areas in a state of arrested succession."

Over the next few years, the pervasive effects of lianas on forests became increasingly clear. In 2005 and 2006, Geertje van der Heijden, then a Ph.D. student working under Phillips, turned to a set of forest plots in Peru to compare the growth rates of hundreds of trees that were or were not infested with lianas. She found that the trees with lianas seemed to grow more slowly and absorb less CO_2, primarily because the vines blocked their sunlight. Van der Heijden estimated that across a single hectare (about 1.4 soccer fields), lianas probably prevented trees from absorbing about 2,000 pounds (920 kilograms) of CO_2 per year.

The biomass of the lianas themselves compensated for 29 percent of that lost CO_2 storage by trees—but they do a much worse job of storing carbon. If you look at the whole forest, says van der Heijden, "lianas are driving a shift toward more carbon [stored] in leaves and less carbon in stems." And while woody stems can lock carbon away for many decades, leaves fall to the ground and decay, returning their carbon to the atmosphere within a few months.

In 2012, van der Heijden teamed up with Schnitzer to do an actual experiment, testing whether lianas really reduce the forest's ability to absorb and store CO_2. A kilometer south of Barro Colorado Island, on the shore of Lake Gatun, workers had used machetes and branch cutters to clear lianas from several swaths of forest. They then returned to record tree growth and death in areas with and without lianas. The results, published in 2015, were dramatic: Even when the biomass of lianas was added in, these vines reduced the forests' overall absorption of CO_2 by 76 percent—or about 8,920 kilograms (19,660 pounds) per hectare per year, nearly ten times what her previous work suggested. "We were quite amazed," says van der Heijden, who is now a forest ecologist at the University of Nottingham.

Despite these discoveries, the causes of liana expansion remained elusive. Schnitzer's team found that higher CO_2 levels boosted liana and tree seedling growth equally, ruling out that explanation. They found the same for nutrients deposited by fossil fuel air pollution. Schnitzer and others recognized that they would need to understand how trees are experiencing the warming climate on an intimate level to get closer to an explanation. A plant physiologist at STRI named Martijn Slot turned to the topic in earnest.

During the early months of 2016, Slot spent many days at the Parque Nacional San Lorenzo, a rainforest preserve on the Caribbean coast of Panama, about 15 kilometers northeast of Barro Colorado. In the dim morning light, Slot would step from the spongy forest floor onto the steel grating of a gondola attached to a twelve-story crane. The contraption hoisted him 30 meters up, past layer after layer of dense foliage; past rainbow-billed toucans that croaked as they flitted away; past an occasional dangling sloth; until he emerged into sunlight among the trees' crowns.

Speaking Spanish into his radio, Slot directed the crane operator through a series of delicate maneuvers that brought his gondola into the fragrant embrace of a leafy branch.

For the next few hours, Slot clamped a small plastic chamber over one leaf after another to measure CO_2 uptake, in order to determine each leaf's rate of photosynthesis. Sometimes an approaching thunderstorm forced him to descend. Other times, his gondola swung in a gust of wind and he accidentally snapped the leaf from its branch, forcing him to start over.

After dozens of crane sessions and 1,700 leaves, Slot found that warm temperatures push trees beyond their comfort zone, even on seemingly mild days. While the air temperature usually stayed below 32°C (91°F), sunlit leaves often surpassed 40°C (104°F). The leaves' photosynthetic rates peaked at about 30°C (86°F) and then plummeted, falling by 90 to 98 percent when the leaf exceeded 40°C (104°F). Slot surmised that this happened because the trees had to triage their competing needs.

During photosynthesis, leaves absorb CO_2 through tiny vents on their undersides called stomata, after the Greek word for "mouth," because they resemble tiny lips. While a leaf's stomata are open, "a huge amount of water is evaporating," says Slot. "For every molecule of CO_2 that's being fixed, there's 300 molecules of water being lost." In this way, a single large tree can exhale around 150,000 liters (40,000 gallons) of water per year.

As the temperature climbs, so too does the rate of evaporation. At some point the leaf closes its stomata—its mouths—and holds its breath. This stops water loss, but also halts photosynthesis, because the leaf can no longer inhale CO_2. Leaves regularly stop photosynthesis during the hottest hours of the day. But rising temperatures could force them to do this more frequently and for longer periods, reducing the trees' food supply.

Many tropical trees "are already operating at their peak temperature," says Kenneth Feeley, a tropical forest ecologist at the University of Miami in Florida. "As it gets hotter, they're becoming less efficient and it's going to lead to slower growth." Amazonia has warmed by 0.9°C (1.6°F) since 1950, with another 1 to 4°C (1.8 to 7.2°F) of warming predicted by 2100, depending on the volume of greenhouse gasses that humans emit. Even if rising CO_2 levels have caused forests to grow faster, as Phillips has found, rising temperatures could eventually erode and reverse the growth-enhancing effects of CO_2. Meanwhile, changes in rainfall are already stressing the Amazon rainforest. The four-month dry season has lengthened by roughly twenty days since 1980, and the region experienced major droughts related to heat waves in 2005, 2010, and 2015–16, killing billions of trees.

This nexus between drought and heat could further boost lianas as trees falter, both now and in the future. Indeed, Schnitzer has found that lianas in Amazonia are more abundant in locales with longer and drier dry seasons. And he and van der Heijden have found that lianas on the south shore of Gatun Lake grow roughly three to four times as quickly as trees during the four-month dry season, even though they grow at similar rates in wet times. These differences especially stood out during the unusually hot and dry 2015–16 El Niño drought, when trees stopped growing entirely, but lianas continued to grow at normal rates. If that happens every five to seven years, whenever there's an El Niño, says Schnitzer, it will give lianas "a huge advantage."

The vines also have other abilities that may compound this one. In November 2015, Medina-Vega, Schnitzer's former postdoc, began an experiment at the Parque Natural Metropolitano, a forest preserve bordering Panama City, which has a pronounced dry season, similar to Barro Colorado Island. There, he selected several dozen tree and liana branches, and meticulously labeled each of their 6,861 leaves with a sharpie. Over the next seventeen months, he measured the growth of each branch and recorded every leaf that fell or sprouted. He found that at this site, lianas searched for sunlight more efficiently than trees did. The skinny liana branches lengthened by up to 38 meters—about fifteen times more than any tree branch did. Liana leaves, like the stems, were also thinner than those of trees, making them "cheap" and "easily replaceable," says Medina-Vega. When a liana's leaf ended up in the shade, the liana simply shed it and replaced it with a new one, somewhere in sunlight. Having cheap leaves also allowed lianas to respond more nimbly to the seasons. Both lianas and trees lose their leaves during the dry season in this location, as in many places in the tropics, but Medina-Vega found that when the rains returned, lianas grew theirs back about a month earlier than the trees did.

So not only do lianas siphon water more quickly than trees, they also punch above their weight in the fight for sunlight. In forests with strong dry seasons, they account for less than 5 percent of the stem biomass—but produce 15 to 40 percent of the leaf mass. Climate change could amplify this advantage. "Let's say these forests are getting drier and drier," says Medina-Vega. "The performance of lianas will improve." And tree growth could suffer.

Despite all the evidence of liana increase, it's still uncertain what it means. Forest changes involve multiple complicated factors and can take centuries to play out, and the lives of human observers are short by comparison.

Tropical ecologist Flavia Costa of the Instituto Nacional de Pesquisas da Amazônia in Manaus believes that site history may have more to do with liana findings than any global phenomenon. Over a ten-year survey of the Adolpho Ducke Reserve in northern Brazil, she and her colleagues found that overall liana abundance remained the same, even though numbers at different plots fluctuated. "I think that data is not as strong as people try to picture," she says.

Barro Colorado's increase, for example, could have come from the fact that it didn't become an island until 1914, when engineers building the Panama Canal dammed a nearby river. The rising waters led to the formation of Gatun Lake, and Barro Colorado, once a hill surrounded by swampland, became isolated at its center. This isolation could have triggered a longterm ecological cascade, if say, populations of large animals such as peccaries or howler monkeys declined as a result, limiting their dispersal of large tree seeds and giving small-seeded lianas a vacuum to fill.

Of course the forest plots that Phillips first used to demonstrate an increase in lianas, in 2002, were far more diverse, including dozens of sites across Peru, Bolivia, Ecuador, Venezuela, Brazil, Costa Rica, and other countries. But individual sites could still have biases. Forest changes occurring today might not be a response to present-day climate change, but rather the gradual drift of a forest back toward an equilibrium after some other event, like a severe windstorm that felled trees a century or two before. "It's inherently hard to pull those apart," admits Phillips. "None of us were around 100 years ago, 200 years ago, to see what the forest was like." Researchers can only hope that working across a broad set of forest sites will cancel these local biases.

Perhaps the biggest mystery, though, is that there's no consistent evidence that lianas are increasing in Africa or Asia. This might come down to a scarcity of long-term data, says Phillips, since scientists have established very few plots to track lianas on these other continents. He speculates that it might also reflect differences in tropical forests across the globe. The few remaining Asian forests often have higher canopies—up to 70 meters, versus 40 or 50 meters in the Amazon—and this might pose a bigger barrier for lianas trying to climb their way to sunlight. African forests, meanwhile, generally sit at higher elevations, making them cooler and damper than most of the Amazon. These forests have milder dry seasons and slower tree turnover, which might blunt lianas' competitive advantages. But if those factors are indeed acting as buffers, they may not help for much longer.

The trees in African forests are starting to show wear and tear from climate change and the consequences could be widespread. Tropical forests play a critical role in helping to offset industrial society's wild overproduction of CO_2. The warning signs that they are increasingly unable to do so have been emerging for some time.

In the years after Phillips's 1994 discoveries about forests turning over more quickly, the University of Leeds ecologist continued to follow the intimate lives of those trees as though they were human subjects in a long-term medical study. Feeding the girth of each tree's trunk into a mathematical equation, his team estimated the total mass of trunk, branches, and leaves. From there, they calculated how much carbon it, and by extension, the surrounding forest, locked away as it grew. This led to a major finding in 2015. Phillips and his University of Leeds colleague Roel Brienen reported that the Amazon's ability to store CO_2 was gradually declining.

Analyzing 321 plots spread across the region, they found that the net amount of CO_2 that the Amazon stored had peaked during the 1990s, at around 2 billion tons of CO_2 per year—roughly the annual emissions of Russia from burning fossil fuels. By the 2000s, it had shrunk by nearly 30 percent.

Phillips sees increasing tree turnover as the biggest driver. "If you stimulate a mature forest, you can probably pack in more trees for a while," he says. "But sooner or later the mortality rate will be going up." He points out that trees which grow faster, or which grow tall at an earlier age, also die sooner. Tall trees suffer more during drought because it's harder for them to suck water all the way up to their leaves. The trunks of faster-growing trees may also be less dense, increasing the risk that wind or the fall of a neighboring tree will break them. And as rising temperatures increasingly interfere with photosynthesis in the future, this could also drive up mortality rates. Those dying trees—like fallen dryads yielding up their souls—will send their carbon wafting back into the air as they decay.

Perhaps most alarming of all, this pattern of reduced CO_2 storage is starting to spread, with African tropical forests also showing early signs. The amount they locked away appears to have peaked around 2010, and is now declining. By 2040, African forests may lose around 50 percent of their annual CO_2 absorption, and, as *bioGraphic* reported in 2017, the Amazon could cross over from absorbing CO_2 to releasing it instead, a recent analysis published in *Nature* predicts. As warming and tree turnover continue to increase, Schnitzer believes that lianas could begin to boom in African forests, too. His hunch stems from his team's 2021 analysis of the liana and tree data that Bernal Vargas and coworkers gathered at Barro Colorado Island from 2016 to 2017. In addition to showing a 29 percent increase in lianas, it also revealed a striking pattern across the landscape.

Nearly all of the new vines had sprung up in places where a falling tree had torn a hole in the canopy—creating an island of sunlight where its fallen lianas re-rooted and sent dozens of new stems slithering up neighboring trees. In other words, the increase in lianas appeared to be driven by the increased turnover in the trees themselves, though the study left open the possibility that the vines are also benefiting from longer dry seasons.

"No study has ever linked the liana increase to any clear mechanistic explanation, and here we do that," says Schnitzer. "This is a massive finding." It might be tempting to view lianas as passive beneficiaries of increased tree turnover, but Schnitzer says they play a more active role. "They're killing the trees and then they're exploiting the gaps that the trees leave," he says. And as those gaps become more closely spaced in the forest—as lianas overrun those gaps and slow the regrowth of trees—a positive feedback loop could take hold. Liana-filled gaps could become the beachheads from which hordes of rapidly growing vines pick and pry their way into swaths of unbroken forest that aren't yet heavily infested.

The critical question is how the liana explosion will impact these forests' ability to absorb CO_2 at a much larger scale. To answer it, van der Heijden, the forest ecologist at Nottingham, plans to expand beyond the research plots and monitor liana growth from space. She and other researchers have spent several years analyzing satellite and aerial images of tropical forests from Central America, South America, and Asia, in hopes of finding ways to estimate the abundance of lianas remotely. If this approach works, then surveys that currently happen only once every few years could occur several times per year, and over thousands of square miles. They could also finally resolve questions about whether liana increase is widespread or simply the result of local biases. So far, the results are tantalizing.

When van der Heijden and her colleagues analyzed images from the Sentinel-2 satellite they found that a hectare of forest shines a bit more greenly when it's infested with lianas, even when that entire hectare—hundreds of trees and lianas—is compressed into a few pixels. It's early days, and van der Heijden and her colleagues are still trying to figure out what the satellites are actually measuring. It may be a true difference in leaf color, or the fact that liana leaves hang more horizontally than tree leaves, or the fact that most liana leaves are thinner than tree leaves, allowing more light to pass through.

Whatever the answer, reaching for satellites seems the most natural extension of the path that Putz blazed forty-plus years ago. Lianas cannot be fully understood from the ground; but viewing them from above, one can see how these vines venture far from their rooted stems, smothering one tree after another as they explore the canopy. Lianas "are the walking plants of the rainforest," van der Heijden says. "They are the ones that can wander."

ALEX CUADROS

Has the Amazon Reached
Its "Tipping Point"?

FROM *New York Times Magazine*

ONE OF THE first times Luciana Vanni Gatti tried to collect Amazonian air she got so woozy that she couldn't even operate the controls. An atmospheric chemist, she wanted to measure the concentration of carbon high above the rainforest. To obtain her samples she had to train bush pilots at obscure air-taxi businesses. The discomfort began as she waited on the tarmac, holding one door open against the wind to keep the tiny cockpit from turning into an oven in the equatorial sun. When at last they took off, they rose precipitously, and every time they plunged into a cloud, the plane seemed to be, in Gatti's words, *sambando*—dancing the samba. Then the air temperature dipped below freezing, and her sweat turned cold.

Not that it was all bad. As the frenetic port of Manaus receded, the canopy spread out below like a shaggy carpet, immaculate green except for the pink and yellow blooms of ipê trees, and it was one of those moments—increasingly rare in Gatti's experience—when you could pretend that nature had no final border, and the Amazon looked like what it somehow still was, the world's largest rainforest.

The Amazon has been called "the lungs of the Earth" because of the amount of carbon dioxide it absorbs—according to most estimates, around half a billion tons per year. The problem, scientifically speaking, is that these estimates have always depended on a series of extrapolations. Some researchers use satellites to detect changes that indicate the presence of greenhouse gases. But the method is indirect, and clouds can contaminate the results. Others start with individual tree measurements in plots scattered across the region, which allows them to calculate the so-called biomass in each trunk, which, in turn, allows them to work out how much carbon is being stored by the ecosystem as a whole. But it's hard to know how representative small study areas are, because the Amazon is almost as large as the contiguous United States, with regional differences in rainfall, temperature, flora and the extent of logging and agriculture. (One study even warned of the risk of "majestic-forest selection bias.")

Gatti's solution was to measure the carbon in the air directly. Which led to the least pleasant part of the flight. The pilot had removed the plane's back seats to make up for the weight of a special silver "suitcase" donated by the U.S. National Oceanic and Atmospheric Administration. Inside, a thick layer of foam cradled seventeen glass flasks with valves that opened and closed at the flick of a switch. Each one was supposed to capture a liter and a half of air from a different altitude, starting at 14,500 feet and going down to 1,000. To ensure that collection always took place above the same point on the map, the pilot had to descend in tight spirals, banking so hard that the horizon went near-vertical.

In a healthy rainforest, the concentration of carbon should decline as you approach the canopy from above, because trees are drawing the element out of the atmosphere and turning it into wood through photosynthesis. In 2010, when Gatti started running two flights a month at each of four different spots in the Brazilian Amazon, she expected to confirm this. But her samples showed the opposite: At lower altitudes, the ratio of carbon *increased*. This suggested that emissions from the slashing and burning of trees— the preferred method for clearing fields in the Amazon—were actually exceeding the forest's capacity to absorb carbon. At first Gatti was sure it was an anomaly caused by a passing drought. But the trend not only persisted into wetter years; it intensified.

For a while Gatti simply refused to believe her own data. She even became depressed. She had always felt a deep connection to nature. As a kid in a distant town called Cafelândia, she would climb a tree in front of her house, spending hours in a formation of branches that seemed custom-made to cradle her arms, legs and head. In later years, no matter how many times she flew over the

Amazon, she never got used to the sight of freshly paved highways, new dirt roads always branching off them, forming a fish-bone pattern. Sometimes she soared past columns of beige smoke that rose all the way to the stratosphere.

Back at her laboratory, which is now housed at Brazil's National Institute for Space Research (INPE), Gatti ultimately spent two years refining her methodology. She wanted to know just how much carbon the rainforest was losing—and even more important, how representative these results were. The whole point of her project was that, by capturing air from such high altitudes, it could provide an empirical and comprehensive picture of the Amazon's so-called carbon budget. So she worked up seven different ways to calculate the effect of wind flows and the composition of air from over the Atlantic Ocean, gradually perfecting her method for subtracting the background noise. Finally she felt confident that her "regions of influence" captured what was happening across 80 percent of the Amazon. The net emissions averaged nearly 300 million tons of carbon per year—roughly the emissions of the entire nation of France.

When Gatti published her findings in *Nature* in 2021, it sparked panicked headlines across the world: The lungs of the Earth are exhaling greenhouse gases. But her discovery was actually much more alarming than that. Because burning trees release a high proportion of carbon monoxide, she could separate these emissions from the total. And in the southeastern Amazon, air samples still showed net emissions, suggesting that the ecosystem itself could be releasing more carbon than it absorbed, thanks in part to decomposing plant matter—or in Gatti's words, "effectively dying more than growing." The first time I spoke to Gatti, she repurposed a lyric by the Brazilian crooner Jorge Ben Jor. How could this be happening, she asked, in a "tropical country, blessed by God / and beautiful by nature"?

The Amazon is a labyrinth of a thousand rivers. They are born at 21,000 feet, with seasonal melts from the Sajama ice cap in Bolivia, and they are born in the dark rock of Peru's Apacheta cliff, as glacial seepage spraying white from its pores. They are born less than 100 miles from the Pacific Ocean; they are born in the middle of the South American continent, in Brazil's high plains, savannas and sandstone ridges. Most are just tributaries of tributaries, headwaters for much larger rivers—the Caquetá, the

Madre de Dios, the Iriri, the Tapajós—any of which, on its own, would already be among the largest rivers in the world. Where these tributaries empty, just south of the Equator, they form the aorta of the Amazon proper, more than 10 miles wide at its widest point. From the Amazon's farthest source to its mouth in the Atlantic, water flows for 4,000 miles, almost as long as the Nile. Measured by the volume it releases into the ocean—the equivalent of a dozen Mississippis, one-fifth of all the fresh water that reaches the world's seas—the Amazon is the largest river in the world.

The consensus used to be that ecosystems are merely a product of prevailing weather patterns. But in the 1970s, the Brazilian researcher Eneas Salati proved that the Amazon, with its roughly 400 billion trees, also creates its own weather. On an average day, a single large tree releases more than 100 gallons of water as vapor. This not only lowers the air temperature through evaporative cooling; as Salati discovered by tracking oxygen isotopes in rainwater samples, it also gives rise to "flying rivers"—rain clouds that recycle the forest's own moisture five or six times, ultimately generating as much as 45 percent of its total precipitation. By creating the conditions for a continental swath of evergreens, this process is crucial to the Amazon's role as a global "sink" for carbon.

Many scientists now fear, however, that this virtuous cycle is breaking down. Just in the past half-century, 17 percent of the Amazon—an area larger than Texas—has been converted to croplands or cattle pasture. Less forest means less recycled rain, less vapor to cool the air, less of a canopy to shield against sunlight. Under drier, hotter conditions, even the lushest of Amazonian trees will shed leaves to save water, inhibiting photosynthesis—a feedback loop that is only exacerbated by global warming. According to the Brazilian Earth system scientist Carlos Nobre, if deforestation reaches 20 to 25 percent of the original area, the flying rivers will weaken enough that a rainforest simply will not be able to survive in most of the Amazon Basin. Instead it will collapse into scrubby savanna, possibly in a matter of decades.

Much of the evidence for this theory—including Gatti's air-sample studies—emerged thanks to a groundbreaking initiative led by Nobre himself. When Nobre started trying to forecast the impact of deforestation back in 1988, he had to do it at the University of Maryland, because his home country lacked the computing power for serious climate modeling. Brazil was so strapped for resources that foreign researchers even dominated Amazon fieldwork. But Nobre spearheaded a program that, in the words of a *Nature* editorial, "revolutionized understanding of the Amazon rainforest and its role in the Earth system." Established in 1999 and known as the Large-Scale Biosphere-Atmosphere Experiment in Amazonia, or LBA, it united disciplines that usually did not collaborate, bringing together chemists like Gatti with biologists and meteorologists. While funding mostly came from the United States and Europe, Nobre insisted that South Americans play leading roles, thus giving rise to a whole new generation of Brazilian climate scientists.

Until recently, Nobre was working under the assumption that the Amazon would not become a net source of carbon for at least another few decades. But Gatti's research is not the only sign that, as he put it to me over Skype, "we are on the eve of this tipping point." The rain machine is slowing. Droughts used to come once every couple of decades, with a megadrought every century or two. But just since 1998 there have been five, two of them extreme. The effect is particularly acute in the eastern Amazon, which has already lost a staggering 30 percent of its forest. The dry season there used to be three months long; now it lasts more than four. During the driest months, rainfall has declined by as much as a third in four decades, while average temperatures have risen by as much as 3.1 degrees Celsius—triple the annual increase for the world as a whole in the fossil-fuel era. In some parts, jungles are already being colonized by grasses.

Losing the Amazon, one of the most biodiverse ecosystems on Earth, would be catastrophic for the tens of thousands of species that make their home there. Rising temperatures could also drive millions of people in the region to become climate refugees. And it would represent a more symbolic death, too, as "saving the rainforest" has long been a kind of synecdoche for modern environmentalism as a whole. What scientists are most concerned about, though, is the potential for this regional, ecological tipping point to produce knock-on effects in the global climate. Because the Amazon's flying rivers circulate back over the continent, the impact may already be reaching beyond the rainforest. In 2015, Brazil's populous southeast was hit by historic water shortages; in 2021, quasi-biblical sandstorms swept the region. If the flying rivers peter out entirely, it could affect atmospheric circulation even beyond South America, possibly influencing the weather as far away as the western United States.

But even these consequences pale next to the fallout from putting the Amazon's carbon back into the atmosphere. For all the slashing and burning of recent years, the ecosystem still stores about 120 billion tons of carbon in its trunks, branches, vines and soil—the equivalent of about ten years of human emissions. If all of that carbon is released, it could warm the planet by as much as 0.3 degrees Celsius. According to the Princeton ecologist Stephen Pacala, this alone would probably make the Paris Agreement—the international accord to limit warming since preindustrial times to 2 degrees—"impossible to achieve." Which, in turn, may mean that other climate tipping points are breached around the world. As the British scientist Tim Lenton put it to me, "The Amazon feeds back to everything."

In May I joined Gatti on a trip to the northeastern Amazon. Though it was not exactly part of her research, she wanted to visit the Tapajós National Forest, a 1.4-million-acre conservation area that held clues to the rainforest's mysterious emissions, and to the transformation predicted by Nobre. First she flew from São Paulo 1,500 miles north to Belém, at the Amazon's mouth in the Atlantic. From there she flew to Santarém, 400 miles upstream, where the Amazon's muddy brown waters are met by the dark blue Tapajós River. In the dry season, tourists come from across Brazil to the Tapajós's white-sand beaches. Now it was raining heavily, the beaches under water. The river lapped at Santarém's sidewalks.

Santarém is one of Brazil's oldest cities, founded by Jesuit missionaries at a time when the only local commodity was Indigenous souls. Its fortunes rose with the rubber boom of the nineteenth century and fell with the bust of the twentieth. More recently, it has been transformed by China's growing demand for soybeans, which are used as livestock feed and cooking oil. Gatti pointed out the long narrow barges docking at a terminal run by Cargill, the American commodities-trading giant. It began operating in 2003, the year before Gatti started running flights from Santarém's tiny airport. As we drove south on the BR-163, also known as Brazil's "grain corridor," Gatti recalled how, back then, so many of the fields were grass for grazing cattle. Of all the deforested land in the Amazon, more than two-thirds is pasture. Here, though, Gatti watched as the grass gave way to a "sea of soy."

Before our trip, Nobre had warned me to keep a low profile, because Gatti had become a public face amid the buzz around her discoveries. Just a few weeks later, the Indigenous-rights advocate Bruno Pereira and the environmental journalist Dom Phillips would be murdered. Profit makes its own law in the Amazon. In the Tapajós region, landowners must preserve 50 percent of their property as rainforest. But Gatti noticed how farmers and ranchers continued to expand their fields, ever so gradually, in long, thin strips meant to evade detection by the satellites of her own employer, INPE. In 2006 the soy industry agreed not to plant in newly deforested areas. But there are ways around this, too. Some farmers bribe local officials for falsified documents. Others transfer land to front men so that they can violate the moratorium without sullying their name. As we drove, Gatti noted violations to report, even though one of her own former colleagues once received death threats for this. She did not hide her affinities, favoring T-shirts with toucans and macaws on florid backgrounds.

Gatti, now sixty-two, has always had a rebellious streak. When she was in college in the late 1970s, some fellow students were arrested for protesting the dictatorship. Outraged, she joined an underground political party and stopped attending classes for a while. Though she was scarcely aware of this at the time, it was the military regime that oversaw the first modern effort to colonize the rainforest. One of its most ambitious projects was the Trans-Amazonian Highway, which pierced 2,600 miles west from the coast and now forms the southern border of the Tapajós National Forest. The goal was partly to fill what the generals saw as a "demographic void," keeping foreign powers like the United States from moving in. They also hoped to relieve pressure on ballooning cities by uniting "men without land in the northeast and land without men in the Amazon." Never mind that the forest was already occupied by a multitude of Indigenous groups; they, too, would be made into productive citizens.

The military regime had also built the BR-163, which branches off from the Trans-Amazonian, forming the Tapajós's eastern border. As we sped along it, signs advertised land for sale, a store called House of Seeds, a World Church of the Power of God. To our right, the Tapajós was a looming wall of green. To our left were private lands where forests were interspersed with croplands. It was the tail end of the soy harvest now, when many landowners started a rotation of corn; tractors rolled through, long metal wings spraying pesticides. Gatti pointed out a freshly cleared area; the trunks lay scattered like a game of pickup sticks. Even when landowners followed the law, what was once a seamless ecosystem became an archipelago, fragments of forest hemmed in by flat expanses. At one point we passed a lone Brazil nut tree, inanely protected by Brazilian law even amid the monoculture. "Here lies the forest," Gatti declared.

As she spoke, Gatti gesticulated so vehemently that both hands sometimes came off the steering wheel. She betrayed no affection for Jair Bolsonaro, the former army officer who spent four years as president pushing to develop the Amazon.

Claiming (baselessly) that his own government's deforestation numbers were a lie, he strangled INPE's funding to the point that it reportedly had to shut off its supercomputer. He also slashed budgets for protecting Indigenous people and the environment. Predictably, deforestation accelerated; in 2021, a thousand trees were cut down every minute. Gatti sometimes thought about quitting, moving with her German shepherd to an eco-villa in the countryside. With Luiz Inácio Lula da Silva back in the presidency, though, she is feeling hopeful for the first time in years. The last time he was in office, from 2003 to 2011, deforestation fell by two-thirds—and now he has promised to halt deforestation entirely. The question is whether that will be enough to halt a process that may now have a momentum of its own.

Eventually Gatti pulled off to the right, through a tunnel of overhanging branches and into an open area where tall trees shaded a research base built as part of Nobre's LBA. The base resembled an eco-lodge, with low-slung wooden buildings topped by clay-tiled roofs. Night was falling, the roar of frogs competing with the distant howl of monkeys. We were met by a thirty-nine-year-old biologist, Erika Berenguer, who wore an old white T-shirt, overlarge and dirty. Her specialty, she said, was *desgraça*—calamity. It turns out that deforestation numbers actually understate the problem of the Amazon, because a fifth of the standing forest has been "degraded" by logging, burning and fragmentation. Now based in Oxford, Berenguer has spent the last twelve years studying how these ills affect the Amazon's ability to store carbon. As she would explain, though, even she was shocked by what happened in 2015, a critical turning point in the health of the ecosystem.

At the time Berenguer's project was to measure every single tree in a few dozen plots in and around the Tapajós National Forest, at regular intervals, to calculate the weight of all the organic matter, or biomass, which serves as a proxy for carbon. At first, when she noticed flames inside the conservation area, she just kept doing her work—gathering up leaf litter, fixing tape around centuries-old trunks, tagging each one with numbered scraps of metal sliced from beer cans. As Berenguer's colleague Jos Barlow likes to point out, outside observers usually fail to distinguish between defor-estation fires (intentionally set to clear freshly clear-cut areas) and wildfires (when the flames accidentally spread to standing forests).

Now it was August, the height of the dry season, when ranchers and farmers in the Amazon clear fields with fire. Almost every year, embers floated across the BR-163 highway, igniting leaves on the forest floor. But the forest itself remained so damp that the flames could not spread far.

Berenguer, a native of cosmopolitan Rio de Janeiro, made a point of sweating alongside her assistants, local men with nicknames like Xarope (Syrup) and Graveto (Stick), whose families had settled by the BR-163 as part of the colonization push of the 1970s. They were not too concerned, either. As subsistence farmers, they also used fire to maintain their lands. It is a tradition that dates back to the region's oldest inhabitants, Indigenous people who discov-ered that ash fertilizes the nutrient-poor soils. Outside the rarest of megadroughts, they never had to worry about losing control of the flames. Researchers have found areas of the Amazon that, according to sediment core samples, went 4,000 years without a single burn.

As Berenguer worked through September, however, the smoke from disparate fires coagulated into a permanent, indistinguish-able haze. It permeated everything—their truck, their clothes, even Berenguer's bra. When they kicked away dead leaves, they noticed that the soil beneath was cracking. The little plants of the understory wilted. Soon everyone was coughing; people took turns breathing mist from a nebulizer, and her own snot turned black. Each morning, she and her assistants had to clean a layer of fresh soot from the windshield of their truck. They turned the brights on, turned the emergency lights on and edged onto the highway. They drove slowly but couldn't see vehicles ahead until they were nearly colliding with them. The sky was hidden. The sun was a red suggestion. Ash fell like alien snow.

The fires were escaping to crop gardens, to pastureland where cattle grazed, to the thatched roofs of houses. And the fires were doing what they should not: spreading inside the rainforest. Splitting her time between Britain and the Amazon, Berenguer had come to know her research plots as intimately as her old neighborhood in Rio. She thought of her favorite places as rainforest versions of her local coffee shop, her local bakery. There were the fallen logs where she and her assistants returned day after day so they could sit and eat lunch. There were the tall, thin buttress roots that acted as a makeshift bathroom stall, hiding her from view when necessary. In one plot, a thick loop of liana hung from the canopy, making for the perfect swing. Now she wanted to save these places.

Among the great old trees of the Tapajós, the flames rose a mere foot from the ground. Berenguer and Xarope could stamp them out with their boots. But their efforts were in vain. The flames consolidated into a thin, uninterrupted arc that stretched for miles into the forest. It advanced slowly, a thousand feet per day; in its wake, the rich perennial green was left brown and gray and charcoal-black. Berenguer watched as animals fled from the fire line—butterflies, deer, thumbnail-size frogs. One day she surprised a snake. It leaped onto a smoldering trunk, accidentally immolating itself, and Berenguer heard a sizzling sound, like buttered bread hitting a griddle plate.

Across the Amazon, more forests ultimately burned than in the largest California wildfires in history, putting half a billion tons of carbon back into the atmosphere—the equivalent of more than one year of emissions by Mexico. It was the Amazon's worst wildfire season on record. Subsequent years have not been as dry, but wildfires have mostly remained well above the average of previous seasons—yet another sign that the ecosystem is losing its natural resilience, entering an alternate feedback loop. In Gatti's samples, the 2015–16 drought also marked the moment when, as she put it to me, "the southeastern Amazon went to pot," and the forest itself started consistently releasing more carbon than it absorbed. Fire does more than destroy trees. It also accelerates the transformations predicted by Nobre's tipping-point theory.

Just about every researcher I spoke to for this article was careful to emphasize their deep respect for Nobre, who has done so much to advance Amazon climate science. But some have reservations about his theory. Partly this is because his earliest simulations showed that, with less rain, the Amazon would give way to the Cerrado, a savanna that covers much of central Brazil. The Cerrado, though, is a carbon-rich patchwork of grasslands, marshes and forests that is itself endangered by global warming and expanding agriculture. How could such a vibrant ecosystem represent ecological collapse? Other researchers, having studied the Amazon up close in mucky fieldwork, object to the use of computer models that apply uniform assumptions to this multifarious biome. Still others express a more pragmatic concern—that the way Nobre communicates his theory is demobilizing. "Carlos gives the impression that the entire forest is going to collapse at the same time, water will stop circulating and it will all become a big savanna," Berenguer told me. Gatti's article, she added, actually led to some misunderstanding, too. Attending the United Nations Climate Change Conference in Glasgow in 2021, she even heard people say that if the Amazon was now a net emitter, why bother saving it?

Nobre himself is aware of these qualms. Now he hastens to clarify that the transformation will take different forms in different regions, and that any end state will be more of an impoverished scrubland than a Cerrado-style savanna. He also predicts that the Amazon's western forests, which are rainier throughout the year because of their proximity to the Andes Mountains, would survive a tipping point. His theory, though, is no longer confined to computer simulations; in the southeastern basin, it may already be playing out. In one study, a team led by the researcher Paulo Brando intentionally set a series of fires in swaths of forest abutted by an inactive soy plantation. After a second burn, coincidentally during a drought year, one plot lost nearly a third of its canopy cover, and African grasses—imported species commonly used in cattle pasture—moved in. Brando also participated in an observational study, led by his colleague Divino Silvério, of the region's enormous Xingu Indigenous Park. Indigenous lands are home to much of Brazil's best-preserved rainforest. But after repeated wildfires, the Xingu's grasslands—traditionally maintained as a source of thatch—nearly tripled in size in less than two decades, to 8 percent of the total area. In the central Amazon, meanwhile, naturally occurring white-sand savannas are taking over seasonally flooded forests—again, largely thanks to fire.

It is tempting to think of climate change as a process that, absent human emissions, would only happen gradually. But as Tim Lenton points out, our planet is naturally prone to "threshold behavior." In a widely cited 2008 paper, Lenton brought the catchy language of tipping points to the arcane revelations of Earth systems science and paleoclimatology. Throughout our planet's history, in individual ecosystems as well as the wider climate, small, incremental changes have started to reinforce one another until—sometimes suddenly— one feedback loop was replaced with a radically different one. What Lenton calls the most "iconic" examples are the Dansgaard-Oeschger events of the last glacial period, when temperatures in Greenland repeatedly shot up by as much as 15 degrees Celsius in the span of a few decades, before cooling again. The causes are intensely debated but most likely involved changes in ice-sheet coverage and the circulation of seawaters.

There is already evidence that our current era of global warming is shifting the borders of various biomes. In Alaska, for example, white spruce trees are moving into areas of tundra for the first time in thousands of years. But humans may have triggered ecological "regime shifts" even before the fossil-fuel era. The Australian Outback was probably lush and green until around 40,000 years ago, when people hunted grass-eating megafauna to extinction, leaving more fuel for fires, which apparently disrupted the continent's own "flying rivers." On Mexico's Yucatán Peninsula, deforestation is thought to have amplified the drought that toppled the Maya. Then there is the Sahara. Ten thousand years ago, the area resembled temperate South Africa, but livestock grazing may have helped turn it into a desert. As the NOAA scientist Elena Shevliakova, who has modeled the global impacts of Amazon deforestation, put it to me, "If a green Sahara is possible, why not a savanna in the Amazon?"

The Amazon has survived ice ages. It may not survive humans. By hastening the demise of its flying rivers, cattle ranchers and soy farmers may be endangering their own livelihoods too. But thanks to what climatologists call teleconnections—weather anomalies linked across thousands of miles—they also threaten agriculture much farther afield. In the El Niño teleconnection, an unusually warm Pacific Ocean pulls the jet stream south, bringing drier conditions to Canada and the northern United States (as well as to the Amazon region). According to a study led by the Notre Dame researcher David Medvigy, a similar pattern could emerge if the Amazon stops recycling its own moisture, as the dry air would travel north in winter. This could halve the snowpack in the Sierra Nevada, a crucial source of water for an already-drought-stricken California.

A growing number of scientists worry that one tipping point can trigger another. In some cases the influence is direct. If Greenland's ice sheet disappears, the circulation of Atlantic seawaters could be drastically altered, which would, in turn, wreak havoc on weather patterns across the globe, making Scandinavia uninhabitably cold, warming the Southern Hemisphere, drying out forests. The impact of Amazon dieback would be to release tens of billions of tons of carbon into the atmosphere—which is more diffuse, but no less dangerous. When Lenton and his colleague David Armstrong McKay recently compiled the latest evidence on an array of global climate thresholds, they found that even a very optimistic 1.5 degrees of warming since preindustrial times may be enough to trigger the gradual but irreversible melting of ice sheets in Greenland and West Antarctica, and to thaw methane-trapping permafrost.

It is difficult to predict how all these shifts might interact, as most models assume, for example, that Atlantic seawaters will always circulate according to known patterns. But in a 2018 paper, Lenton and the American Earth system scientist Will Steffen warned that a dominolike "tipping cascade" could push the global climate itself beyond a critical threshold, into an alternate feedback loop called "hothouse Earth," with hostile conditions not seen for millions of years. It can feel like doom-mongering to contemplate such a scenario. There is no way to put a number on it. Even if it is improbable, however, Lenton argues that the consequences would be so dire that it must be taken seriously. He sees it as a "profound risk-management problem": If we focus only on the most likely outcomes, we will never predict anomalies like 2021's unprecedented "heat dome" in the Pacific Northwest. Or last year's winter heat wave in Antarctica, when temperatures jumped 70 degrees Fahrenheit above the average. Or, for that matter, the proliferation of wildfires in the world's largest rainforest.

Berenguer wanted to show Gatti how the 2015 megafires had altered forests in the northeastern Amazon. So Xarope picked us up from the research base in the morning, and we got back onto the BR-163. Here and there along the highway, Berenguer pointed out "tree skeletons"—dead trees whose sun-bleached branches poked from the otherwise lush green canopy of the Tapajós. Fire did not always kill them right away. When Berenguer was back in Britain, her assistants would send her updates by WhatsApp. *You know Tree 71?* one message might say, referring to a centuries-old specimen in one of her plots. *So, it just died.* It could take a few years more for it to fall to the ground. Some of the carbon in Gatti's air samples, then, could be a delayed consequence of past fires. But as we would see inside the living forest, something stranger was happening, too.

Eventually we exited the highway for an unmarked dirt track that ended in a wall of vegetation. Machete in hand, Berenguer led us onto a tight path. Just a few days earlier, she and her assistants spent hours clearing the way for us, but new vines were already reclaiming the space. "You can see it's a mess," Berenguer said. An impassable thicket of reedy bamboo hemmed us in on either side; the canopy was low above our heads. To me it looked normal enough, as far as jungle goes. In reality, though, a healthy rainforest should be easy to walk through, because the largest trees consume so much light and water that the understory lacks the resources to grow very dense.

We walked over fallen trunks. Unlike in the southeastern Amazon, Berenguer still saw no evidence of savanna-like vegetation moving in. But the balance of native species was now out of whack, as opportunistic "pioneers" occupied the spaces left by dead giants. In some areas, fast-growing embaúba trees stood so uniformly that they resembled the stems of a wood-pulp plantation. In others, hundreds of newborn lianas formed a kind of snake nest. (Berenguer's team had to measure each one individually, a hellish task.) She pointed to a tall, proud tree that had somehow survived the blaze. Because all of the other nearby individuals of its species had been killed, it was unlikely to reproduce; Berenguer called it a "zombie."

A University of Birmingham researcher named Adriane Esquivel-Muelbert has found similar changes across the Amazon. Even in the absence of actual "savannization," trees that can withstand drier conditions are proliferating, while those that need more water are dying in greater numbers. The dominance of embaúba is particularly worrisome because the trees are hollow, storing far less carbon than a slower-growing species like mahogany. Their life cycle is also relatively short, leaving more frequent gaps in the canopy. The end result of this transformation is unclear, but Gatti's numbers have only continued to get worse. According to her latest five-year averages, the Brazilian Amazon is already giving off 50 percent more carbon than it was in the first five years of her project—and even the historically healthier western forests are sometimes emitting more than they absorb.

Eventually we came into a clearing. I began to sweat. The sun was searing hot; Berenguer said that unshaded ground can reach 176 degrees Fahrenheit here. Clearings are a natural part of the Amazonian cycle, as large trees inevitably die and other species gradually take their place. But even logging could not match the power of fire to turn the forest into "Swiss cheese." Berenguer never used to need sunscreen because the canopy was so thick; now she gets sunburned here. And the profusion of holes sets off a vicious cycle. The sun dries out the vegetation; trees shed leaves to preserve water; the litter becomes fuel for the next fire. The gaps also create a "wind corridor," allowing strong drafts to penetrate deep into the forest during storms. Perversely, with their heavy trunks, the largest, oldest trees are especially vulnerable to being knocked over.

"This used to be a beautiful forest," Xarope said.

"Some days it makes me sad," Berenguer said. "Other days it pisses me off. This is one of those days."

Berenguer had hoped that the misfortune of the megafire would at least provide an opportunity to study how a rainforest recovers from such *desgraça*. But she worried that she would never find out, because it would never get the chance. Among scientists who study the Amazon, the notion of multiple tipping points, specific to each region's ecology, has increasingly taken hold. And some now speak of an even more urgent "flammability tipping point," past which an ecosystem that never evolved to burn starts burning regularly. During the drought of 2015, wildfires also ravaged another nearby conservation area, the Reserva Extrativista Tapajós-Arapiúns. Because it was left so degraded, with so much dried-out fuel on the floor, there was a much more intense conflagration in 2017, even though that was a wet year. This time the flames were not the foot-tall ones that are usually seen in the Amazon but reached all the way to the canopy.

Though her mainstay was ecological calamity, Berenguer also

wanted us to see what well-preserved old-growth forest looks like. In strictly scientific terms, it was a control, a necessary point of comparison with the messed-up forests, as she called them (though she used a more colorful word that cannot be printed here).

She also let on that she welcomed the rare excuse to traipse around a more "David Attenborough" setting. So we drove south on the BR-163 until we hit the 117th kilometer marker, where we re-entered the Tapajós.

We were walking for only a few minutes before the difference became obvious. It was cooler and darker. The flora was far more varied, forming distinct layers as you lifted your eyes to the sky. The canopy was far more closed, the understory far more open; Berenguer and Xarope didn't even need to prune the trail for our visit. There were lianas here, too, but they were few and large. One was as thick as a tree; Berenguer said it was probably centuries old.

It's hard to shake a popular image of scientists as rigorously rational, unemotional about their work. But Berenguer was not embarrassed to admit that, as she put it, she and her colleagues have their own personal tipping points, too. For a while after the 2015 fires, she lost her sense of purpose, the hope that her work could make a difference. The flames had even ravaged the plot where she used to swing on that perfect loop of liana. "Your whole reference system is being destroyed, and you're powerless," she said. "It's hard to explain without sounding like a tree-hugger. Not to say I don't hug trees, because I do." Some trees were too big for that, though. Here was an urucurana, with its winglike buttress roots taller than my whole body. Here was a soaring strangler fig, which surrounds another tree's trunk as it grows, eventually killing its host. "What a dirty trick!" Gatti exclaimed.

At one point we came upon a low tree bearing a yellow fruit that neither Berenguer nor Xarope could identify.

"Is it poisonous?" Gatti asked.

"I don't know," Xarope said. Then he plucked one from a branch and bit into it. We did the same. There was not much pulp around the stone but the flavor was sharp and rich.

Berenguer remembered a past research trip to track frugivores—fruit-eating creatures. She and her colleagues had to remain absolutely still and silent for hours to avoid spooking them. I suggested we try it out for a minute, just to hear what an old-growth forest sounds like without humans tramping around.

We stopped walking; Berenguer sat on a log. As our chitchat faded, the racket of birds swelled as if someone had suddenly turned the volume dial on a stereo. I closed my eyes for a moment. When I looked again, Berenguer's eyes had narrowed to slits, her lips curled into a faint smile. Earlier, describing what she felt in this place, Berenguer used the word *grandeza*, which literally means greatness, but also bigness. The rainforest made her feel small, and she liked this.

Gatti had spoken about feeling, at least temporarily, not so separate from the natural world—almost as if she were a kid again, ensconced in that tree in front of her house. Now she stood with her eyes shut, palms open at her sides as if she were at a religious revival, as if she were receiving something.

I glanced over at Xarope; he looked amused. Then the spell was broken by the more familiar sound, distant but unmistakable, of a semitruck shifting gears.

AMANDA GEFTER

What Plants Are Saying About Us

FROM *Nautilus*

I WAS NEVER into house plants until I bought one on a whim—a prayer plant, it was called, a lush, leafy thing with painterly green spots and ribs of bright red veins. The night I brought it home I heard a rustling in my room. Had something scurried? A mouse? Three jumpy nights passed before I realized what was happening: The plant was *moving*. During the days, its leaves would splay flat, sunbathing, but at night they'd clamber over one another to stand at attention, their stems steadily rising as the leaves turned vertical, like hands in prayer.

"Who knew plants do stuff?" I marveled. Suddenly plants seemed more interesting. When the pandemic hit, I brought more of them home, just to add some life to the place, and then there were more, and more still, until the ratio of plants to household surfaces bordered on deranged. Bushwhacking through my apartment, I worried whether the plants were getting enough water, or too much water, or the right kind of light—or, in the case of a giant carnivorous pitcher plant hanging from the ceiling, whether I was leaving enough fish food in its traps. But what never occurred to me, not even once, was to wonder what the plants were thinking.

I was, according to Paco Calvo, guilty of "plant blindness." Calvo, who runs the Minimal Intelligence Lab at the University of Murcia in Spain where he studies plant behavior, says that to be plant blind is to fail to see plants for what they really are: cognitive organisms endowed with memories, perceptions, and feelings, capable of learning from the past and anticipating the future, able to sense and experience the world.

It's easy to dismiss such claims because they fly in the face of our leading theory of cognitive science. That theory goes by names like "cognitivism," "computationalism," or "representational theory of mind." It says, in short, the mind is in the head. Cognition boils down to the firings of neurons in our brains.

And plants don't have brains.

"When I open up a plant, where could intelligence reside?" Calvo says. "That's framing the problem from the wrong perspective. Maybe that's not how our intelligence works, either. Maybe it's not in our heads. If the stuff that plants do deserves the label 'cognitive,' then so be it. Let's rethink our whole theoretical framework."

Calvo wasn't into plants, either. Not at first. As a philosopher, he was busy trying to understand human minds. When he began studying cognitive science in the 1990s, the dominant view was the brain was a kind of computer. Just as computers represent data in transistors, which can be in "on" or "off" states corresponding to 0s and 1s, brains were thought to represent data in the states of their neurons, which could be "on" or "off" depending on whether they fire. Computers manipulate their representations according to logical rules, or algorithms, and brains, by analogy, were believed to do the same.

But Calvo wasn't convinced. Computers are good at logic, at carrying out long, precise calculations—not exactly humanity's shining skill. Humans are good at something else: noticing patterns, intuiting, functioning in the face of ambiguity, error, and noise. While a computer's reasoning is only as good as the data you feed it, a human can intuit a lot from just a few vague hints—a skill that surely helped on the savannah when we had to recognize a tiger hiding in the bushes from just a few broken stripes. "My hunch was that there was something really wrong, something deeply distorted about the very idea that cognition had to do with manipulating symbols or following rules," Calvo says.

Calvo went to the University of California San Diego to work on artificial neural networks. Rather than dealing in symbols and algorithms, neural networks represent data in large webs of associations, where one wrong digit doesn't matter so long as more of them are right, and from a few sketchy clues—*stripe, rustle, orange, eye*—the network can bootstrap a half-decent guess—*tiger!*

Artificial neural networks have led to breakthroughs in machine learning and big data, but they still seemed, to Calvo, a far cry from living intelligence. Programmers train the neural networks, telling them when they're right and when they're wrong, whereas living systems figure things out for themselves, and with small amounts of data to boot. A computer has to see, say, a million pictures of cats before it can recognize one, and even then all it takes to trip up the algorithm is a shadow. Meanwhile, you show a two-year-old human *one* cat, cast all the shadows you want, and the toddler will recognize that kitty.

"Artificial systems give us nice metaphors," Calvo says. "But what we can model with artificial systems is not genuine cognition. Biological systems are doing something entirely different."

Calvo was determined to find out what that was, to get at the essence of how real biological systems perceive, think, imagine, and learn. Humans share a long evolutionary history with other forms of life, other forms of mind, so why not start with the most basic living systems and work from the bottom up? "If you study systems that look way different and yet you find similarities," Calvo says, "maybe you can put your finger on what is truly at stake."

So Calvo traded neural networks for a green thumb. To understand how human minds work, he was going to start with plants.

It turns out it's true: Plants do stuff.

For one thing, they can sense their surroundings. Plants have photoreceptors that respond to different wavelengths of light, allowing them to differentiate not only brightness but color. Tiny grains of starch in organelles called amyloplasts shift around in response to gravity, so the plants know which way is up. Chemical receptors detect odor molecules; mechanoreceptors respond to touch; the stress and strain of specific cells track the plant's own ever-changing shape, while the deformation of others monitors outside forces, like wind. Plants can sense humidity, nutrients, competition, predators, microorganisms, magnetic fields, salt, and temperature, and can track how all of those things are changing over time. They watch for meaningful trends—Is the soil depleting?

Is the salt content rising?—then alter their growth and behavior through gene expression to compensate.

Plants' abilities to sense and respond to their surroundings lead to what seems like intelligent behavior. Their roots can avoid obstacles. They can distinguish self from non-self, stranger from kin. If a plant finds itself in a crowd, it will invest resources in vertical growth to remain in light; if nutrients are on the decline, it will opt for root expansion instead. Leaves munched on by insects send electrochemical signals to warn the rest of the foliage, and they're quicker to react to threats if they've encountered them in the past. Plants chat among themselves and with other species. They release volatile organic compounds with a lexicon, Calvo says, of more than 1,700 "words"—allowing them to shout things that a human might translate as "caterpillar incoming" or "*$@#, lawn mower!"

Their behavior isn't merely reactive—plants anticipate, too. They can turn their leaves in the direction of the sun before it rises, and accurately trace its location in the sky even when they're kept in the dark. They can predict, based on prior experience, when pollinators are most likely to show up and time their pollen production accordingly. A plant's form is a record of its history. Its cells—shaped by experience—remember.

Chat? Anticipate? Remember? It's tempting to tame all those words with scare quotes, as if they can't mean for plants what they mean for us. For plants, we say, it's biochemistry, just physiology and brute mechanics—as if that's not true for us, too.

Besides, Calvo says, plant behavior can't be reduced to mere reflexes. Plants don't react to stimuli in predetermined ways—they'd never have made it this far, evolutionarily speaking, if they did. Having to deal with a changing environment while being rooted to one spot means having to set priorities, strike compromises, change course on the fly.

Consider stomata: tiny pores on the undersides of leaves. When the pores are open, carbon dioxide floods in—that's good, that's breathing—but water vapor can escape. So how open should the stomata be at any given time? It depends on the availability of water in the soil—if there's plenty more for the taking, it's worth letting the carbon dioxide in. If the dirt's dry, the leaves have to retain water. For the leaves to make that decision, the roots have to tell them about the availability of water. The leaves communicate their own needs to the roots in turn, encouraging them, for example, to form symbiotic relationships with specific microorganisms in the soil.

If a plant could respond to sensory information on a one-to-one basis—when the light does *x*, the plant does *y*—it would be fair to think of plants as mere automatons, operating without thought, without a point of view. But in real life, that's never the case. Like all organisms, plants are immersed in dynamic, precarious environments, forced to confront problems with no clear solutions, betting their lives as they go. "A biological system is never exposed to just a single source of stimulation," Calvo says. "It always has to make a compromise among different things. It needs some kind of valence, a higher-level perspective. And that's the entry to sentience."

Sentience?

Are plants clever? Maybe. Adaptive? Sure. But sentient? Aware? *Conscious?* Listen closely and you can hear the scoffing.

To feel alive, to have a subjective experience of your surroundings, to be an organism whose lights are on and someone's home—that's reserved for creatures with brains, or so says traditional cognitive science. Only brains, the theory goes, can encode mental representations, models *of* the world that brains experience *as* the world. As Jon Mallatt, a biologist at the University of Washington, and colleagues put it in their 2021 critique of Calvo's work, "Debunking a Myth: Plant Consciousness," to be conscious requires "experiencing a mental image or representation of the sensed world," which brainless plants have no means of doing.

But for Calvo, that's exactly the point. If the representational theory of the mind says that plants can't perform intelligent, cognitive behaviors, and the evidence shows that plants *do* perform intelligent, cognitive behaviors, maybe it's time to rethink the theory. "We have plants doing amazing things and they have no neurons," he says. "So maybe we should question the very premise that neurons are needed for cognition at all."

The idea that the mind is in the brain comes to us from Descartes. The seventeenth-century philosopher invented our modern notion of consciousness and confined it to the interior of the skull. He saw the mind and brain as separate substances, but with no direct access to the world. The mind was reliant on the brain to encode and represent the world or conjure up its best guess as to what the world might be based on, with ambiguous clues trickling in through unreliable senses. What Descartes called "cerebral impressions" are today's "mental representations." As cognitive scientist Ezequiel Di Paolo writes, "Western philosophical tradition since Descartes has been haunted by a pervasive mediational epistemology: the widespread assumption that one cannot have knowledge of what is *outside* oneself except through the ideas one has *inside* oneself."

Modern cognitive science traded Descartes' mind-body dualism for brain-body dualism: The body is necessary for breathing, eating, and staying alive, but it's the brain alone, in its dark, silent sanctuary, that perceives, feels, and thinks. The idea that consciousness is *in* the brain is so ingrained in our science, in our everyday speech, even in popular culture that it seems almost beyond question. "We just don't even notice that we are adopting a view that is still a hypothesis," says Louise Barrett, a biologist at the University of Lethbridge in Canada who studies cognition in humans and other primates.

Barrett, like Calvo, is one of an increasing number of scientists and philosophers questioning that hypothesis because it doesn't comport with a biological understanding of living organisms. "We need to get away from thinking of ourselves as machines," Barrett says. "That metaphor is getting in the way of understanding living, wild cognition."

Instead, Barrett and Calvo draw from a set of ideas referred to as "4E cognitive science," an umbrella term for a bunch of theories that all happen to start with the letter *E*. Embodied, embedded, extended, and enactive cognition—what they have in common (besides *E*s) is a rejection of cognition as a purely brainbound affair. Calvo is also inspired by a fifth *E*: ecological psychology, a kindred spirit to the canonical four. It's a theory of how we perceive without using internal representations.

In the standard story of how vision works, it's the brain that does the heavy lifting of creating a visual scene. It has to, the story goes, because the eyes contribute so little information. In a given visual fixation, the pattern of light in focus on the retina amounts to a two-dimensional area the size of a thumbnail at arm's length. And yet we have the impression of being immersed in a rich three-dimensional scene. So it must be that the brain "fills in" the missing pieces, making inferences from scant data and offering up its best hallucination for who-knows-who to "see," who-knows-how.

Dating back to the work of psychologist James Gibson in the

1960s, ecological psychology offers a different story. In real life, it says, we never deal with static images. Our eyes are always moving, darting back and forth in tiny bursts called saccades so quick we don't even notice. Our heads move, too, as do our bodies through space, so what we're confronted with is never a fixed pattern of light but what Gibson called an "optic flow."

To "see," according to ecological psychology, is not to form a picture of the world in your head. It stresses that patterns of light on the retina change relative to your movements. It's not the brain that sees, but the whole animate body. The result of "seeing" is never a final image for an internal mind to contemplate in its secret lair, but an adaptive, ongoing engagement with the world.

Plants don't have eyes exactly, but flows of light and energy impinge on their senses and transform in predictable ways relative to the plants' own movements. Of course, to notice that, you first have to notice that plants move.

"If you think that plants are sessile," or stationary, Calvo says, "just sitting there, taking life as it comes, it's difficult to visualize the idea that they are generating these flows."

Plants appear sessile to us only because they move slowly. Quick movements—like the nightly shuffle of my prayer plant—can be accomplished by altering the water content in certain cells to change the tension in a stem, or to stiffen a branch under the weight of heavy snow. Most plant movement, though, occurs through growth. Since they can't pick up their roots and walk away, plants change location by growing in a new direction. We humans are basically stuck with the shape of our bodies, but at least we can move around; plants can't move around, but they can grow into whatever shape best suits them. This "phenotypic plasticity," as it's called, is why it's critical for plants to be able to plan ahead.

"If you spend all this time growing a tendril in a particular direction," Barrett says, "you can't afford to get it wrong. That's why prediction does seem very important. It's like my granddad said; maybe all granddads say this: 'measure twice, cut once.'"

Phenotypic plasticity is a powerful but slow process—to see it, you have to speed it up. So Calvo makes time-lapse recordings, in which slow and seemingly random growth blooms into what appears to be purposeful behavior. One of his time-lapse videos shows a climbing bean growing in search of a pole. The vine circles aimlessly as it grows. Hours are compressed into minutes. But when the plant senses a pole, everything changes: It pulls itself back, like a fisherman casting a line, then flings itself straight for the pole and makes a grab.

"Once movement becomes conspicuous by speeding it up," Calvo says, "you see that certainly plants are generating flows with their movement."

By using these flows to guide their movements, plants accomplish all kinds of feats, such as "shade avoidance"—steering clear of over-populated areas where there's too much competition for photosynthesis. Plants, Calvo explains, absorb red light but reflect far-red light. As a plant grows in a given direction, it can watch how the ratio of red to far-red light varies and change directions if it finds itself heading for a crowd.

"They are not storing an image of their surroundings to make computations," Calvo says. "They're not making a map of the vicinity and plotting where the competition is and then deciding to grow the other way. They just use the environment around them."

That might seem to be a long cry from how humans perceive the world—but according to 4E cognition, the same principles apply. Humans don't perceive the world by forming internal images either. Perception, for the *E*s, is a form of sensorimotor coordination. We learn the sensory consequences of our movements, which in turn shapes how we move.

Just watch an outfielder catch a fly ball. Standard cognitive science would say the athlete's brain computes the ball's projectile motion and predicts where it's going to land. Then the brain tells the body what to do, the mere output of a cognitive process that took place entirely inside the head. If all that were true, the player could just make a beeline to that spot—running in a straight line, no need to watch the ball—and catch.

But that's not what outfielders do. Instead, they move their bodies, constantly shuffling back and forth and watching how the position of the ball changes as they move. They do this because if they can keep the ball's speed steady in their field of vision—canceling out the ball's acceleration with their own—they and the ball will end up in the same spot. The player doesn't have to solve differential equations on a mental model—the movement of her body relative to the ball solves the problem for her in active engagement, in real time. As the MIT roboticist Rodney Brooks wrote in a landmark 1991 paper, "Intelligence Without Representation,"

"Explicit representations and models of the world simply get in the way. It turns out to be better to use the world as its own model."

If cognition is embodied, extended, embedded, enactive, and ecological, then what we call the mind is not in the brain. It is the body's active engagement with the world, made not of neural firings alone but of sensorimotor loops that run through the brain, body, and environment. In other words, the mind is not in the head. Calvo likes to quote the psychologist William Mace: "Ask not what's inside your head, but what your head's inside of."

When I first encountered the 4E theories, I couldn't help thinking of consciousness. If the mind is embodied, extended, embedded, etcetera, does consciousness—that magical, misty stuff—seep out of the confines of the skull, permeate the body, pour like smoke from the ears, and leak out into the world? But then I realized that way of thinking was a hangover from the traditional view, where consciousness was treated as a noun, as something that could be located in a particular place.

"Cognition is not something that plants—or indeed animals—can possibly *have*," Calvo writes in his new book, *Planta Sapiens*. "It is rather something created by the interaction between an organism and its environment. Don't think of what's going on inside the organism, but rather how the organism *couples* to its surroundings, for that is where experience is created."

The mind, in that sense, is better understood as a verb. As the philosopher Alva Noë, who works in embodied cognition, puts it, "Consciousness isn't something that happens inside us: It is something we do."

And we do it in order to keep on living. The need to stay alive, to tread in far-from-equilibrium water—*that* is what separates us from machines. "Wild cognition," as Barrett puts it, is more akin to a candle flame than to a computer. "We are ongoing processes resisting the second law of thermodynamics," she says. We are candles desperately working to re-light ourselves, while entropy does its damnedest to blow us out. Machines are made—one and done—but living things make themselves, and they have to remake themselves so long as they want to keep living.

The Chilean biologists Humberto Maturana and Francisco Varela—founding fathers of embodied and enactive cognition—coined the term "autopoiesis" to capture this property of self-creation. A cell—the fundamental unit of life—serves as the prime example.

Cells consist of metabolic networks that churn out the very components of those networks, including the cell membrane, which the network continuously builds and rebuilds, while the membrane, in turn, allows the network to function without oozing back into the world. To keep its metabolism going, the cell needs to be in constant exchange with its environment, drawing in resources and tossing out waste, which means the membrane has to let things pass through it. But it can't do it indiscriminately. The cell has to take a stance on the world, to view it as a place of value, full of things that are "good" and "bad," "useful" and "harmful," where such terms are never absolute but dependent on the cell's ever-changing needs and the environment's ever-changing dynamics.

These valences, Calvo says, are the stirrings of sentience. They are distinctions that carve out (or "enact") a world in a process that 4E cognitive scientists call "sense-making." The act of making valenced distinctions in the world, which allow you to draw the boundary between self and other, is the primordial cognitive act from which all higher levels of cognition ultimately derive. The same act that keeps a living system living is the act by which, as Noë puts it, "the world shows up for us."

"You start with life," says Evan Thompson, a philosopher at the University of British Columbia and one of the founders of the enactive approach. "Being alive means being organized in a certain way. You're organized to have a certain autonomy, and that immediately carves out a world or a domain of relevance." Thompson calls this "life-mind continuity." Or as Calvo puts it, echoing the nineteenth-century psychologist Wilhelm Wundt, "Where there is life there is already mind."

From a 4E perspective, minds come before brains. Brains come into the picture when you have multicellular, mobile organisms—not to represent the world or give rise to consciousness, but to forge connections between sensory and motor systems so that the organism can act as a singular whole and move through its environment in ways that keep its flame lit.

"The brain fundamentally is a life regulation organ," Thompson says. "In that sense, it's like the heart or the kidney. When you have animal life, it's crucially dependent for the regulation of the body, its maintenance, and all its behavioral capacities. The brain is facilitating what the organism does. Words like 'cognition,' 'memory,' 'attention,' or 'consciousness'—those words for me are properly applied to the whole organism. It's the whole organism that's conscious, not the brain that's conscious. It's the whole organism that attends or remembers. The brain makes animal cognition possible, it facilitates and enables it, but it's not the *location* of it."

A bird needs wings to fly, Thompson says, but the flight is not *in* the wings. Disembodied wings in a vat could never fly—it's the whole bird, in interaction with the air currents shaped by its own movements, that takes to the sky.

"Plants are a different strategy of multicellularity than animals," Thompson says. They don't have brains, but according to Calvo they have something just as good: complex vascular systems, with networks of connections arranged in layers not unlike a mammalian cortex. In the root apex—a small region in the tip of a plant's root—sensory and motor signals are integrated through electrochemical activity using molecules similar to the neurotransmitters in our brains, with plant cells firing off action potentials similar to a neuron's, only slower. Like the human brain, the root apex allows the plant to integrate all of its sensory flows in order to produce new behavior that will generate new flows in ways that keep the plant adaptively coupled to the world.

The similar roles played by an animal's nervous system and a plant's vascular system help explain why the same anesthetics can put both animals and plants to sleep, as Calvo demonstrated using a Venus flytrap in a bell jar. Normally, the plant's traps snap shut when an unfortunate insect triggers one of its sensor hairs, which protrude from the trap's mouth like sharks' teeth. (Actually, the clever plant awaits the triggering of a second hair within seconds of the first before expending the costly energy to bite. Once closed, it awaits three more triggers—to ensure there's a decent bug buzzing around in there—before it releases acidic enzymes to digest its meal. As Calvo sums it up, "They can count to five!") Using surface electrodes, Calvo watched as the triggered hairs sent electric spikes zapping through the plant, sparking its motor system to react. With anesthesia, all of that stopped. Calvo tickled the trap's hairs and it just sat there, its mouth agape. The electrode reading flatlined.

"The anesthesia prevents the cell from firing an action potential," Calvo explains. "That happens in both plants and animals." It's not that the anesthetic is turning down the dial of consciousness inside the brain or root apex, it's just severing the links between sensory inputs and motor outputs, preventing the organism from engaging as a singular whole with its environment. Once "woken," though, the groggy Venus flytraps quickly returned to their usual behavior.

"Clearly," Thompson says, "plants are self-organizing, self-maintaining, self-regulating, highly adaptive, they engage in complex signaling among each other, within species and across species, and they do that within a framework of multicellularity that's different from animal life but exhibits all the same things: autonomy, intelligence, adaptivity, sense-making." From a 4E perspective, Thompson says, "there's no problem in talking about plant cognition."

In the end, Calvo's critics are right: Plants aren't using brains to form internal representations. They have no private, conscious worlds locked up inside them. But according to 4E cognitive science, neither do we.

"The mistake was to think that cognition was in the head," Calvo says. "It belongs to the relationship between the organism and its environment."

After talking with Calvo, I looked around my apartment overrun with plants—at the pothos and bromeliads, rocktrumpet vines and staghorn ferns, at the peace lilies and crowns of thorns, snake plants, Monstera, ZZs, and palms—and they suddenly appeared very different. For one thing, Calvo had told me to think of plants as being upside down, with their "heads" plunged into the soil and their limbs and sex organs sticking up and flailing around. Once you look at a plant that way, it's hard to unsee it. But more to the point, the plants appeared to me now not as objects, but as subjects—as living, striving beings trying to make it in the world—and I found myself wondering whether they felt lonely in their pots, or panicked when I forgot to water them, or dizzy when I rotated them on the windowsill.

It wasn't just the plants. I felt myself differently, too: less like a passive spectator, snug inside my skull, and more like an active life form, tendrilled and strange, moving through the world as the world moved through me.

"Plants are not that different from us after all," Calvo had told me, "not because I'm beefing them up to make them more similar to us, but because I'm rethinking what human perception is about. I'm neither inflating them nor deflating us but putting us all on the same page."

It was hard not to wonder whether, from that page, the story of our planet might unfold differently. The *E* approaches ask us to question what we are, how intimately we're entangled with the world, and whether we can rightly see ourselves as standing apart from nature or whether the destruction we wreak is steadily diminishing our own wild cognition.

"Human nature," wrote John Dewey, the pragmatist philosopher, "exists and operates in an environment. And it is not 'in' that environment as coins are in a box, but as a plant is in the sunlight and soil. It is of them."

A Good Prospect

FROM *The Drift*

FOUR BILLION YEARS ago, our planet was a restive place, full of geological commotion. At the earth's center, a molten metal core began to coalesce, while heat and radioactive energy kept large swaths of the surface liquid. Violent volcanic forces made and re-made the landscape. Over eons, magma pooled and hardened, forming some of the oldest and most stable parts of the earth's crust. Heat, pressure, and fluid heavy with chemicals left deposits of metals and minerals in these ancient, crystalline substrates of the planet.

In early March, I found myself examining a small cross section of one such deposit. Whitish gray quartz and feldspar were speckled with a pale, shimmering green silicate called spodumene. For millennia, this particular specimen had reposed deep below ground about 30 miles north of Lake Superior, part of a geologic formation called the Canadian Shield. It sat there until the invisible hand of the global market reached down into the earth, removed it, and transported it to the Metro Toronto Convention Centre, where a friendly geologist named Ramin Ghaderpanah was calling my attention to the spodumene. Those green bits, he told me, can be refined into high-grade lithium. And right now, lithium is one of the most sought-after minerals in the world, making it "such an attractive investment," according to Ghaderpanah, who works for a Canadian mining company with additional lithium projects in Argentina and Nevada. Even as recently as three to four years ago, he said, lithium was not in high demand here. By "here," the geologist was referring to our immediate surroundings: a commotion of excited dealmaking and unrestrained investment taking place over four days at the largest annual mining conference in the world, held in Toronto by the Prospectors & Developers Association of Canada (PDAC).

The geologist and I were standing at one of the more than 1,500 booths that filled the cavernous convention center, which this year welcomed nearly 24,000 attendees representing more than 130 countries. At times, the mining crowd seemed to outnumber the locals, filling the sidewalks beneath the smooth, gleaming high-rises of downtown Toronto. PDAC is where the people, companies, governments, and other institutions that constitute the global metal-mining industry commiserate in the bad times and celebrate the good. The names of the world's mining titans, each worth many billions of dollars, were plastered on every available surface. A massive Newmont banner hung over the escalator. Freeport provided a café and lounge; BHP sponsored admirably fast Wi-Fi.

There were also investors, both large firms and individual traders. All were looking for the right stock—an up-and-coming miner, or an obscure mineral deposit with promise. They spent their days talking to company representatives, geologists, prospectors, and mineral officials from all over the world. In addition to the registered attendees, many more showed up for the informal, but no less important, conference behind the conference—the backroom meetings, the chats in hotel bars, the parties. Perhaps especially the parties. These happened every night, often sponsored by some firm or investor. I spent one late night at a pub called the Walrus, and struggled for a time to find an actual mining employee. Instead, I met a transit technology salesman, a software designer, a few people who work for market makers, and a securities analyst from Australia. All had come to cut deals, schmooze, and see old friends.

It has been several years since the industry felt inclined to let loose. Metal prices took a dive in the mid-2010s, with gold, silver, and copper prices all bottoming out in 2014 and 2015. Mass layoffs ensued. Reuters described the 2015 PDAC as cowed, with fewer parties and empty floor space; the CEO of a gold-mining company described "far less prime rib, far more chips, far more salsa." By the end of the decade, metal prices had recovered; they have now recovered again after a brief, Covid-related slump in 2020. At the height of the pandemic PDAC was held online and then, last year, in hybrid form.

This year, prime rib was back—metaphorically, but also on top of a cracker that I ate at a swank open bar. If PDAC is a weather vane for global mining, this year's event made one thing clear: the industry thinks that the winds of commerce are at its back. Metal miners stand on the verge of a planet-spanning, multi-decade mineral boom, driven by the demands of an electrifying world. Global decarbonization to address climate change will require enormous amounts of graphite and manganese, nickel and cobalt. Above all, it will require copper. Without copper, we cannot build solar panels, wind turbines, or electric cars and their battery chargers. S&P Global, the market research firm, expects copper demand to double by 2035 and climb thereafter, dramatically outstripping supply. "In the 21st century, copper scarcity may emerge as a key destabilizing threat to international security," its 2022 report found.

Lithium is also expected to win big. A world that does not rely on fossil fuel combustion will need rechargeable battery technology at an unprecedented scale, in everything from the cars we drive to grid-scale energy storage infrastructure, with enough capacity to power a city when the sun sets and everyone turns on their lights. None of this will be possible without lithium. The Biden White House estimates that demand for lithium and other electric vehicle (EV) battery minerals could swell by 4,000 percent in the coming decades. And, as with copper, there's evidence that global lithium supply will soon be insufficient—without a production boom, we'll have only half the lithium and cobalt we need to hit 2030 climate goals, according to the International Energy Agency (IEA). A May report by the Carnegie Endowment for International Peace laid out still more dire projections: in 2030, lithium, cobalt, and graphite demand may outpace production for the U.S. and its allies tenfold, thirtyfold, and eightyfold, respectively.

Expected shortages and bottomless demand have automakers scrambling to secure their supply chains. In January, General Motors announced a $650 million investment in Lithium Americas, which is developing what will likely be the U.S.'s largest lithium mine. Since then, G.M. has also invested in a lithium-extraction startup and announced a fourth planned battery manufacturing plant in the U.S. And in the past year, mining giant Rio Tinto and Chinese battery manufacturer CATL both signed deals with Ford.

To call the mood at PDAC optimistic would be an understatement. The conference was awash with talk of new mines and big profits. Amid the hubbub, I caught up with John Thompson, a geoscientist and longtime mining insider who seemed to know everyone. Thompson spoke softly, in a British accent, and projected an ironic detachment from the surrounding commotion, born of many years in the industry. "The buzz," he said, "is related to the perceived importance of critical metals and minerals. You can debate lots of issues around that, but the industry is just feeling ecstatic. Everybody loves us, and that hasn't traditionally been the case."

Mining is getting a makeover. To build a mine, a company needs legal permits, and in the old days, perhaps those were enough. Today, though, the industry and its investors increasingly believe that in order to be successful—and maximize profits—a company also needs what the industry calls a "social license to operate," or moral permission to reap the benefits of tearing up the earth to extract minerals. Social license and profit go hand in hand. With that in mind, companies are trying to reinvent themselves as part of the solution to the climate crisis, allies to the environmentally minded with carbon-neural targets for their global operations. And that's not all: this year's PDAC also displayed a heightened interest in the concerns of Indigenous nations and a focus on increasing the number of women in the industry. Many panels and speeches began with land acknowledgements. Sessions on offer included "Why Indigenous women in mining is a golden opportunity," "The amazing race to decarbonize," and "Operationalizing the 'S' in ESG: Does it matter to investors?" (ESG stands for environmental, social, and governance, a shorthand for a supposedly more socially conscious form of investment.) Unequivocally, the answer was yes.

From the opening keynote address, it was clear that the climate crisis itself has become a means for mining interests to obtain social license, providing a ready justification for the industry's activities. Following a land acknowledgement, Ken Hoffman, head of the battery materials team at McKinsey, took the podium in front of a large room packed with hundreds of people. Hoffman summarized the state of global supply, demand, and pricing trends for various key metals. He discussed the significant material needs of electric vehicles and several renewable energy technologies. An electric car requires six times the amount of mineral resources as a gas-powered car, according to the IEA; an onshore wind turbine outstrips a natural gas–fired power plant in mineral inputs by a factor of nine. He encouraged the industry to position itself to deliver ethically sourced, low-carbon metals and minerals to meet specifications laid out by regulators in the U.S. and E.U., and discussed various advances in EV technology. The upshot of each of these topics, though, was the same. Decarbonizing the modern world is going to make the mining world a lot of money—by Hoffman's estimate, on the order of fifteen to twenty trillion dollars. At one point, Hoffman seemed to address a nameless, climate-conscious consumer, the sort of person who wants their personal choices to reflect their desire to save the planet—a desire that, in all likelihood, will enrich the people in that room.

"To stop global warming," he said, "you need us."

The phrase "energy transition" is a common shorthand for the elimination of fossil fuels. But, as the economic historian Adam Tooze argued in March, its suggestion of a smooth shift from one mode to another fails to adequately capture the radical nature of the challenge ahead—the total transformation of global energy production required to address the climate crisis. Coal, gas, and oil still account for more than 60 percent of humanity's total electricity generation. These need to be phased out immediately; extant and planned fossil fuel projects are almost certain to push the globe past two degrees Celsius of warming. And new energy sources will need to meet surging global electricity demand, which is expected to double, at minimum, in the coming decades. "The wholesale displacement of fossil fuels across global electricity generation, with overall capacity expanded to twice its current size, in the space of a single generation, will be a truly staggering undertaking," Tooze writes.

This monumental economic transformation will require a lot—and I mean a lot—of minerals and metals. As a result, Western governments and corporations are scrambling to secure their mineral supply, keen to enact the necessary changes while also meeting the quality-of-life expectations of first-world consumers. In the U.S., the Inflation Reduction Act (IRA) provides half a trillion dollars for clean-energy and climate policies, which includes tax breaks and incentives for electric cars and battery-metal supply chains. PDAC attendees approve, of course. As Hoffman, the

McKinsey analyst, said, the IRA changed "everything." But it also led to some jokes poking fun at the U.S. as a great consumer of metals with a longstanding aversion to doing the mining or refining itself. "In the U.S. they don't want to mine," Hoffman told the amused crowd, "but they want to buy from sources deemed acceptable to U.S. regulators. Canada, we love you."

All that cash, however, cannot hide the plain fact that the U.S. and other Western governments no longer call all the shots. To obtain battery metals, countries are forming and breaking partnerships, striking deals and making enemies, in ways that mirror the shifting alliances of our increasingly multipolar world. In the Anglophone sphere, governments have closed ranks against China, which refines more than half of the world's lithium and controls some 85 percent of the processing capacity for rare-earth metals. In recent decades, this arrangement seemed to satisfy the scions of global mining, whose headquarters are in the industrial core: Canada, the U.K., Australia, Switzerland. Minerals are often extracted in the Global South, processed in China, and turned into capital back home on a few key stock exchanges. (Australia, London, and Toronto host the primary metal markets. PDAC's location is no accident; most large mining firms are either based in Canada or have large offices there.)

Chafing against its status as a mere stop on the supply chain, China has become a producer and consumer of metals and rare-earth minerals at levels rivaling those of the West. Several Chinese companies challenge the Western heavyweights in market capitalization, with huge operations in places like Tibet, Mongolia, and sub-Saharan Africa. In February, Nigeria's government announced a large lithium deal with one of the major Chinese miners, months after rejecting a bid from Tesla. Fearing competition from the new superpower, Canada forced three Chinese companies to divest from lithium holdings last November. A few months later, Australia blocked an attempt by a Chinese firm to buy a greater share in a mining company. Both governments cited national security concerns.

Now that Western governments' influence is on the wane, countries rich in deposits of battery minerals across the Global South have discovered that they have leverage. More than half of the world's lithium reserves are concentrated in South America's "Lithium Triangle"—Chile, Bolivia, and Argentina—and in the past few years, a wave of left-wing and socialist governments have come to power on the continent. This is not a coincidence. Most of the winning candidates promised some combination of a crack-down on mining pollution and increased respect for the rights and wishes of Indigenous communities, who have historically been forced to bear the costs of mineral extraction. For many of these countries, mineral wealth has meant impoverishment for the peo-ple who live on the richest land. Former Peruvian President Pedro Castillo and President Gabriel Boric of Chile were elected after campaigning on redistributing mining profits and stronger min-ing regulations. In February, Mexican President Andrés Manuel López Obrador signed a decree that placed all the country's lith-ium under state control. And Gustavo Petro, Colombia's furthest-left president in seventy-five years, has seized the assets of two gold companies, proposed huge taxes on the mining sector, and vowed to form a state-owned mining company. In April, *The New Republic* reported on attempts to undermine Petro, backed by right-wing military and business elites.

Naturally, this trend has left companies and investors anxious for friendlier sites to mine. In February, just before PDAC, a prominent mining podcast warned about escalating "jurisdiction risk," noting that investors "hear all sorts of stories about South America, you know with Peru and protests, and all this sort of stuff." This gave the episode's guest, Argentina Lithium & En-ergy CEO Nikolaos Cacos, a chance to tout the stability and pro-mining governance in Argentina, where his company has several large lithium claims (which give a company the right to explore for minerals on a specific parcel of land). "There's elections coming up," he said, "and there's projections for a pro-business government to win." Other global players, it seems, are willing to submit to the new demands of South American governments. German Chancellor Olaf Scholz recently visited Chile, hoping to divert some of its lithium, which predominantly goes to China, to his country's automakers. According to Bloomberg, Scholz pledged to invest in Chilean processing of raw materials, rather than exporting them.

Demands for greater domestic rewards from mineral exploita-tion can be heard far beyond South America. Indonesia banned nickel exports, which has helped promote a fast-growing battery industry, and is trying to organize an OPEC-like consortium for metal processing. In January, the Philippines proposed a 10 percent nickel ore export tax, prompting outcry from private-sector firms, which argued that the move would "kill" the industry, Reuters reported. And a government watchdog from the Democratic Republic of the Congo, home to large copper and cobalt deposits, which state-run Chinese companies have mined for over a decade, recently took the superpower to task for delivering less than a third of a promised three billion dollars for infrastructure improvements. It seems China, wary of losing those deposits, will play ball. According to Reuters, the two are negotiating an additional seventeen billion dollars and a larger stake for the DRC's state-run mining company.

Political power and economics have always been entwined, free trade and international markets backstopped by military might and geopolitical muscle. Still, it seems that, in the frantic race for limited battery-metal supplies, governments are unwilling to leave things to the market, wielding their power in increasingly explicit ways. All the rapid shifts in the global mining industry's geography introduced a current of doubt into PDAC's party atmosphere. The green metals boom is indeed here. Beyond that, uncertainty abounds.

The convention floor at PDAC was overwhelming. There was a booth for a diamond mine in Angola; a booth for Greenland's mining delegation; another for a geochemistry consulting firm. There were advertisements for copper mines from Arizona and the Xanadu copper-and-gold project in Mongolia. The national mining company of Chad offered pamphlets. "The country is virgin from an industrial point of view," they read, "an open market with little competition." There were smiles and handshakes. Everyone was selling hope and excitement and the promise of profit. One booth displayed bars of gold bullion. Some exhibits were clean and modern; others were cluttered with technical maps and dense data that required an advanced degree in geology to decipher.

I passed the Core Shack and the Investors Exchange and found my way to the last row of booths, set against the back wall of the gigantic building. This is where PDAC puts the prospectors, the treasure hunters who scour maps and mineral reports for deposits of ore that the major companies have passed over. They stake a claim, maybe drill a few exploratory core samples, and bring their finds to the conference, hoping that a small mining or exploration company will throw some capital their way. It's "about the riskiest thing you can do for money," an American prospector once told me.

I stopped by one booth, an outfit from the island of Newfoundland, and met Neal Blackmore, who was glad to talk me through the travails of prospecting. "We're pretty low on the food chain," he told me. In some Canadian provinces, staking a mining claim still requires actually driving wooden posts into the ground. With his tangled beard and long hair, a rumpled blazer worn over a graphic t-shirt, Blackmore seemed out of place at the buttoned-up corporate conference. It was easy to imagine him bushwhacking through the woods and pounding stakes.

Still, his look seemed to serve him well at PDAC. While we were talking, two capital markets investors and the CEO of an exploration company stopped by. All were working with him on potential mines, building on deals hammered out, in part, at previous PDACs. He made dinner plans with the men and discussed the group's mining camp in Newfoundland, where the main danger comes in the form of trigger-happy big-game hunters from America. Blackmore's table displayed a few copper-gold sulfide samples, as well as papers describing gold and silver deposits—and a single lithium claim, which a woman scooped up while we were talking. Blackmore is new to lithium. He started learning about it only last year. "We were always gold guys," he said. "We go wherever the market is."

I told Blackmore that it was my first time at PDAC and asked him what I should know. "A mine is a hole in the ground with a liar standing on the edge," he said. His point was not that PDAC was full of scam artists, though the conference's history does include some notable grifts. Backers of the notorious Bre-X fraud amplified a fake gold claim in Indonesia over the course of several PDACs in the 1990s. The company went from trading as a penny stock to reaching six billion Canadian dollars in market value before the sham fell apart. A geologist involved in the scheme allegedly committed suicide by jumping out of a helicopter over the jungle. One investigation concluded that the geologist had been strangled, his suicide faked, according to the *Calgary Herald*. There are persistent rumors that he was later spotted alive, according to a well-connected international-mining source who knew some of the players firsthand. The affair was loosely adapted into a 2016 movie starring Matthew McConaughey.

Blackmore meant that mining is risky. Markets are fickle. And promising ore deposits often prove disappointing. When someone says that a mine or a trend is a sure thing, some skepticism is probably warranted. Just a few years ago, cobalt was the battery metal of the day, but its scorching stock-market rise led to an attendant bust—and now other minerals, like manganese, are replacing it in the supply chain. All of this adds an extra layer of intrigue at PDAC. While the schedule includes numerous technical panels that cater to geologists and data nerds, PDAC is, ultimately, an investment conference. Companies large and small spend four days certifying the quality of their ore deposits, the cozy relationships they have with government regulators, the heights they expect their stock to climb. Investors, meanwhile, are seeking good bets for their money.

These are not the sort of people, in other words, who tend to endorse nascent left-wing governments asking multinational mining companies to pay up. But that's precisely what the leaders of Chile, which has the largest lithium reserves in the world, plan to do. The conflict between Western financial interests and an empowered Global South has rapidly come to define the shape of the battery-metals boom. And it became concrete on PDAC's second morning, at a presentation by Chile's mining delegation. Outside the conference room, the hallway was packed. Men in navy suits soberly conversed in English and Spanish over weak coffee, filling every seat once the doors opened. On a raised dais at the front of the room sat a group of Chilean mineral and diplomatic officials, including Raúl Fernández, the ambassador to Canada, and the country's undersecretary for mining, Willy Kracht.

The Chilean economy is inextricably tied to mining. It is the world's largest copper producer and second-largest lithium producer behind Australia. The industry makes up more than half of Chile's exports and employs one out of every thirteen workers. Kracht, who has a thick black beard and a calm demeanor, is a top mining official in Gabriel Boric's government. A leftist elected in 2021, Boric pledged to overhaul decades of pro-business, low-regulation governance in Chile—especially related to natural resources. Last year, the administration held a constitutional referendum, meant to replace Chile's Pinochet-era governing framework. At one point, it looked like the new constitution was going to nationalize all lithium resources. Voters ultimately rejected the proposal. Kracht told the PDAC crowd that the reforms were overly ambitious; they contained "too much of everything." He reassured investors that future constitutional measures were unlikely to include full nationalization of the sought-after mineral. Going forward, he said, Chile would create a publicly managed lithium company and pursue other reforms, like raising the royalty tax on mining companies and increasing its domestic copper smelting capacity, rather than shipping the concentrate overseas. "We have the right as a country to have ownership in the industry," he told the crowd.

Kracht then turned to the elephant in the room—the "social unrest" that had "people around the world worried." In 2019, protests against public-transit prices, police corruption, and widespread unemployment spread from the capital city of Santiago to every region in Chile. The protests led to the downfall of President Sebastián Piñera, a billionaire businessman with a cozy relationship to the mining industry—the Pandora Papers revealed that Piñera took money from a mining executive in exchange for support of an environmentally destructive copper and iron mine—and propelled Boric into office. Kracht seemed intent on reassuring the crowd that investing in Chile would be worth it. He pulled up a graph that showed foreign financing of Chilean mining had increased since 2018, despite the ongoing protests. "Markets understand," he said.

It was hard to gauge the impact of Kracht's reassurances. Even among their own, in a boom-time atmosphere, PDAC attendees were noticeably wary about making definitive statements about much of anything. At a booth for Albemarle, one of two private companies currently producing Chilean lithium, I asked a representative whether the country's new left-wing government and years of protests worried potential investors. "Yes," she said, but wouldn't elaborate. She directed me to her supervisor, who wasn't there and declined to comment over email. (Kracht's chief of staff suggested that some of his remarks "may be misinterpreted," but didn't dispute any specific claim.)

As it turned out, Kracht's conciliatory tone was a head fake. In April, about six weeks after PDAC, Boric announced in a nationally televised speech that his administration planned to do what PDAC attendees might not have been led to expect: establish state ownership of all Chile's lithium. This transition, per Reuters, will require that private companies either partner with the state, which will command majority ownership, or give up their operations after their mining permits expire. In his speech, Boric referenced Chile's nationalization of copper in 1971 under President Salvador Allende, who was subsequently deposed by Pinochet in a coup supported by the CIA. Public lithium ownership, Boric said, will help create "a Chile that distributes wealth we all generate in a more just way." (Stock prices for lithium-mining companies in the country dipped after this news, as media reports described investors as "spooked.")

This news came after PDAC, but corporate unease about unruly left-wing governments was in evidence at the conference, if you found the right room. One panel, called "Mining related disputes with a focus on Latin America: political risk and mitigation tools," featured representatives from two law firms—one Colombian, the other Ecuadorian. Panelists instructed audience members on how they might use international commerce agreements to protect their investments from leftists. The platform of Colombian President Petro, who has pledged to support Indigenous rights against mining companies, "is based unfortunately on the rights of ethnic communities," said Álvaro José Rodríguez, a Colombian natural resource lawyer. (Rodríguez later elaborated over email, writing, "What I meant when I said that Gustavo Petro's political platform 'is based unfortunately on the rights of ethnic communities' is that the Petro government is too focused on the rights of ethnic communities and less so on the rights of other stakeholders and the needs of the country.")

"He is not only a leftist, but it is fair to call him a populist," Rodríguez said. "He was elected on a pro-minority and pro-environmental platform."

There was at least one person at PDAC free—and by that I mean rich enough—to speak his mind. That was Robert Friedland, a billionaire financier with a knack for funding the exploration of rich and obscure mineral deposits. These include a nickel find in Labrador, which he sold in the 1990s for more than four billion Canadian dollars, and a colossal gold mine in Mongolia. Like so many others, Friedland had something to sell at this year's conference. A PDAC luminary, he was given a headline speaking slot to make his pitch to a packed conference hall. By the time I got there, nearly every chair was claimed. I stood in a crowd at the back, awkwardly taking notes against a pillar.

At seventy-two, Friedland speaks with the meandering cadence of an old hippie, which his biography suggests that he is. In the 1970s, he left Bowdoin College after he and two others were caught with 24,000 tablets of LSD in what was then, according to the *New York Times*, the "largest ever" seizure of the drug in New England. Later, he ran an apple orchard where Steve Jobs took psychedelics (the experience helped inspire the name "Apple"). In his speech, Friedland touched on topics ranging from the steam engine to the Blues Brothers and the Harvard-Yale football game, made fun of Joe Biden, and suggested that the U.S. should not go to war with China over Taiwan because Taiwanese and Chinese people are hard to tell apart. Friedland also touted his new mining technology startup, I-Pulse, with funding from Bill Gates, Jeff Bezos, and mining giant BHP. He lamented the good old days of a more integrated global economy, back before the financial crisis, tariff fights, and trade wars, when China was "making almost everything the world consumed." Friedland was especially unhappy with the political shifts in South America. As examples he listed the "crazy people running Colombia"; the "thirty-six-year-old communist" in Chile; "Evo" (who is no longer the president of Bolivia); and Brazilian president "Lula, a proven crook."

Fortunately for the audience, Friedland has a solution for investors who share his fear of South America's leftward turn: southern Africa, where, it turns out, Friedland is developing some enormous projects with his company, Ivanhoe Mines. He described the region as "relatively unexplored" in its mineral resources and boasting a ready workforce. "The virus bounced off them because they're young," he told the crowd.

Mid-speech, Friedland paused to play videos that stitched together dramatic footage of mining equipment in motion and hydroelectric power generation for net-zero carbon operations, interspersed with shots of smiling local workers, all backed by a pounding EDM soundtrack. The first project he was promoting, Kamoa-Kakula in the DRC, is set to produce more than 500,000 tons per year of the highest grade copper. Here, Ivanhoe is partnering with Zijin Mining Group, one of China's largest companies. The second, which Friedland called the largest precious-metals project in the world, is in South Africa. Beginning next year, the

Platreef mine is expected to yield a treasure trove of sought-after minerals—platinum, palladium, nickel, gold, rhodium.

Friedland is far from alone in turning to Africa to feed the world's infinite appetite for battery metals. In a bid to counter China's tightening hold on southern Africa's mining industry, this past January the U.S. signed a memorandum of understanding to create a battery supply chain with the DRC and Zambia. The DRC supplies about 70 percent of the cobalt used in rechargeable batteries, with approximately a quarter of that coming from what are known as "artisanal small-scale mines," according to a report from the NYU Stern Center for Business and Human Rights. These are often illegal operations that use child laborers. Hundreds of thousands of Congolese artisanal miners dig for cobalt and copper with picks and shovels, handling toxic heavy metals with their bare hands for as little as one dollar per day. Through tangled networks of traders, buyers, and processing firms, these metals end up in the products of the largest companies in the world, including Tesla and General Motors. (In 2022, Tesla's shareholders rejected a proposal that would have required detailed reporting on child labor in its mineral sources.)

Governments and companies descending on the Congo and exploiting its resources: there's a term for it. "This economic model is a colonial model," Jacques Nzumbu, a Congolese Jesuit priest, told me. Nzumbu recently moved to Montreal for a graduate program, but before that he spent years in the Congo working with artisanal miners in Lualaba, a province in the southern DRC that is a significant source of cobalt and copper. Nzumbu described protests—put down by the Congolese army in 2019—against the mining giant Glencore. After a cave-in killed at least forty-three small-scale miners, who had been sneaking into a Glencore-run copper-and-cobalt site to do their own mining, state security forces brutally evicted thousands of other independent miners nearby. "Local people were very angry with Glencore," Nzumbu recalled. "We see in the Congo every day, more than 5,000, 7,000 trucks, with cobalt and copper going out of our country without transformation" of Congolese society.

At PDAC, though, the general sense was that metal mining will be transformative. It will help save the planet and provide jobs, an assumption summed up in the title of the panel "Better lives, better climate: Latin America and its minerals." Mining's necessity was accepted; its toll on people and land was quickly smoothed over, when mentioned at all. Most expressions of environmental concern seemed calculated to justify more mining. Indeed, this was how Friedland ended his speech: with a Carl Sagan quote and a call for the mining industry to save the planet. The slogan of one of Friedland's subsidiary companies, Ivanhoe Electric, drove home the point: "Reinventing mining for the electrification of everything."

If you've seen photos of lithium extraction, they are probably of brine ponds: huge neon-turquoise lakes that stand in striking contrast to the surrounding white-gray desert. Lithium tends to occur in alkaline, mineral-rich brines on salt flats. This water is pumped to the surface, mixed with other chemicals, and left to evaporate in the scorching desert sun, leaving a concentrate behind. Lithium production requires more water than other battery metals, but when it comes to sheer environmental destruction, it's hard to beat copper. Copper is extracted from gigantic open-pit mines, hundreds to thousands of feet deep. To refine copper ore, sulfuric acid solutions are typically leached through the rock piles, which creates toxic slurries that carry mercury, arsenic, and other poisonous metals. At large copper mines—like Bingham Canyon in Utah, the world's deepest open-pit mine—this pollution requires treatment "in perpetuity" to prevent spills and damage to groundwater, according to a report by the environmental nonprofit Earthworks. Every day at the small mines in the Congo, artisanal miners climb into pits and tunnels filled with toxic water from cobalt and copper extraction. Specifics of production vary by mineral type, but spiking demand for battery metals will mean a lot more mining of this sort all over the world.

The mining industry has always provided an economic justification for the displacement and exploitation of people all over the world: the colonization of the Americas for gold, silver, iron, and copper; blood diamonds in West and Central Africa; child labor in cobalt mines in the DRC; and thousands of deaths linked to paramilitaries financed by multinational mining companies in Colombia. The move to renewable energy will likely expose the poorest people on the planet to more of the same from these fierce extractive forces. A *Nature* study published late last year found that more than half of the materials needed for the green energy transformation are located on or near relatively undeveloped land where Indigenous and peasant populations live. As mines encroach on these communities, they will be removed from their homelands or forced to live with profound industrial pollution.

The focus at this year's PDAC was primarily on the economic benefits for these communities. First Nations speakers attested that mining can be good for Indigenous communities. "Before the mine, we had nothing," Donny McCallum told the crowd at a panel on Indigenous economic inclusion. McCallum is a member of the Marcel Colomb First Nation (MCFN), in Manitoba. In 2022, the nation signed joint ventures with two contracting companies designed to help secure employment for members at a nearby gold mine. "We want a piece of the pie," McCallum explained. The panel's moderator, Christian Sinclair, who is Opaskwayak Cree, encouraged Canada's First Nations to follow MCFN's example and form Indigenous economic development corporations. He pointed to the Southern Ute Tribe of southwestern Colorado, which sits atop a rich formation of coal-bed methane. The tribe used these resources to build a three-billion-dollar organization.

The only note of hesitation about tying the well-being of First Nations to mining industry profits was sounded by an older man with a long ponytail during the Q&A portion of this session. "The land cannot sustain making the most amount of money in the least amount of time," he told the panelists.

I caught up with this man, Rick Cheechoo, later on, at a reception for PDAC's Indigenous Program over mini bison potpies and wild rice salad. He is a member of the Moose Cree First Nation. A large gold mine operates near his nation's traditional territory, he told me. There have been benefits—he described agreements between the mining company and the First Nation that provide jobs and healthcare—but a company's drive for profit puts it at odds with some tribal needs, especially preserving their culture and treaty rights and ensuring that families displaced by industry are compensated.

Historically, mining has brought disruption and violence to Canada's First Nations, and there's no reason to believe that's changing. The past decade has seen numerous confrontations, with protestors blockading mines and arrested en masse by militarized police. Even as the conference was happening, members of the Naskapi and Innu nations were fighting an iron mine in

Quebec—a situation that, as far as I could tell, was not addressed by anyone at PDAC.

Are there no alternatives to this rush to extract the world's metals and minerals? At PDAC, almost no one asks this question. The assumption was baked into virtually every convention-floor booth and conference-room panel: minerals must be extracted. The market and a cooler planet demand it. ESG and other signs of virtuous consumption, like partnerships with Indigenous communities, permit it.

Even beyond PDAC, alternate visions can be hard to come by, but a new study from the Climate and Community Project and the University of California, Davis, aims to expand our imaginations. It is an important piece of scholarship, arguing that we may not need to accept a future in which the mining industry—with the blessing of governments—continues to tear up the world's forests and occupy its deserts. Focusing on lithium consumption, the report models different developmental pathways for the U.S., using variables like car ownership, the size of EV batteries, city density, public transit, and battery recycling. The worst-case scenario, the authors claim, would result in major lithium extraction, as PDAC attendees expect. But they show that reducing the size of EV batteries could shrink expected U.S. lithium demand in 2050 by 42 percent. (Car companies, it's worth noting, are not trending toward smaller vehicles. In April, G.M. announced that it was ending production of the Chevy Bolt, the company's smallest EV, in order to build more electric trucks and SUVs.) But even if average battery size were to remain the same, cutting car ownership rates, largely by creating denser cities and better public transit, could shrink total lithium demand by somewhere between 18 percent and 66 percent, according to the study.

The spread, or not, of recycling and reusing minerals is another crucial variable. John Thompson, the longtime industry insider I spoke to, attended PDAC this year in part as a representative of Regeneration, a company that plans to re-mine abandoned sites and use the profits to restore their original ecosystems. He's a recycling proponent, and hopes that the rest of the industry will catch on soon. "Everybody would say recycling is important," he told me. But "most people in this conference aren't interested."

Even the most optimistic version of the future, involving reduced demand and robust recycling, will still require some mining. What this ought to look like increasingly preoccupies Patrick Donnelly, who works for the Center for Biological Diversity in Nevada. I know Donnelly—as do a lot of other journalists in the West who cover extractive industries—as a ferociously dedicated conservation advocate. A few years ago, Donnelly realized that no one was tracking all of the American lithium projects and decided to do so himself. His map now shows more than 115 potential mines, clustered in his home state. "It's the biggest mineral rush of our lifetime," he told me over the phone.

In an ideal world, Donnelly said, the U.S. government would put a moratorium on speculative claims and instead survey all of the country's mineral deposits in order to identify the least harmful places to mine. This isn't happening and won't anytime soon. In May, the U.S. fast-tracked a manganese and zinc mine in Arizona, the first mining project added to a program designed to expedite the clean-energy transition and other infrastructure developments. But Donnelly also fears that anti-mining sentiment is turning people against electric cars—and against lithium extraction altogether. "There is zero chance we can recycle our way out of the problem," he said. This is true. There isn't enough lithium on the market for battery recycling to realistically meet present demand, let alone the expected increase.

"There is an element of the mining resistance movement that opposes not just particular mines but all lithium and all electric vehicles," Donnelly went on. "Unless we're talking about deindustrializing society, which I don't think appeals to most people, we need to be thinking about how and where we're getting our lithium, and critically examine our own use of these minerals, like the cell phone I'm speaking to you on now, with minerals from South America, where locals say the mines are destroying their environment and community."

Such are the paradoxes of the globalized green economy, in which blocking a mine in one place means shifting extraction somewhere else. We want to decarbonize, yet our lives require ever-increasing supplies of energy. And so climate-minded consumers and the mining industry are locked in a self-justifying embrace. We buy an EV and think we are doing right by those vulnerable to rising temperatures and tides. But in trying to continue consuming as we are used to, buying stuff and zipping down the highway, we have exposed many of those same vulnerable people to another threat—the market's readiness to kill, poison, and displace them to get minerals and metals. The mining industry, meanwhile, benefits from the self-satisfied consumerism of the EV buyer. For all of its disdain for environmentalists, the industry needs green consumers who seek absolution for their carbon-intensive ways of life. With their complacent inattention to the injustices inflicted by the green economy, these consumers not only fund the industry's expansion but give it moral cover.

PDAC started with a party and ended with another. After the first full day, I joined a horde of people streaming through the doors of a large conference hall for a networking event. Hundreds of people listened as a speaker gave a brief land acknowledgement, backlit by a facade emitting purple light. The crowd skewed male and young. Undercuts and blazers without ties were the dominant style. Then the real entertainment began, a deafening "dueling pianos" show that ruled out the possibility of my interviewing anyone. The first song: "Sweet Child o' Mine."

The concluding party was an awards gala at the Fairmont Royal York hotel, an elegant art deco building a few blocks from the convention center. Tickets for the affair, which included a three-course meal, were $225 per person, or $2,250 for a table. I declined to buy a ticket, but stopped by the hotel anyway.

I got there just as it was getting dark. Fierce, cold winds blew off Lake Ontario and roared down the skyscraper canyons of downtown Toronto. A small crowd of protestors had congregated outside. They waved signs, prayed, and tried to pass tea-light candles to the mining officials who pushed past them into the lobby. Attached to the candles were small note cards that described water poisoned by a Canadian gold company in Argentina, land defenders murdered by security forces in Guatemala, and the 2019 dam collapse in Brazil, when a barrier holding back iron-mine waste gave way, releasing a massive toxic slurry that inundated a low-lying village and killed 270 people. (Top officials from Vale, a Brazilian company with offices in Toronto, face murder charges.)

Most of the protestors were members of Catholic social justice organizations, with connections to communities in the Global South impacted by mining. They were joined by solidarity organizations tied to the Philippines and Peru. I talked to one of the protestors, Dean Dettloff, as the event wound down. A nearby sign declared "Hands off Africa," a quote from Pope Francis's critique of conflict diamonds and the green-metals boom.

"We need to transition to green energy, but at whose expense?" Dettloff said. "Who is being thrown under the wheels of that transition? Those are the people we want to amplify and be in solidarity with."

Dettloff and other protestors told me they had tried to have conversations with PDAC attendees. I brought up the industry's emphasis on addressing climate change over the past few days. After all, the world cannot abandon fossil fuels without a good deal more lithium, copper, and many more minerals. I told him what the McKinsey analyst had said—"to stop global warming, you need us"—a sentiment that, however opportunistic, contains an uneasy truth. Did it suggest a possible area of shared interest with the industry, a possible path to a conversation? "My response to that would be that I need them to stop harming my friends in the Global South," Dettloff said. "When we have that conversation first, maybe then we can have another conversation about sustainable minerals."

I thanked the protestors and hurried off to the nearest streetcar, head down against the wind. The last of the activists packed up their cardboard coffee containers and signs. The lights from the Fairmont Royal York gave off a golden yellow glow. Inside, the party was going strong.

Hot Air

FROM *The New Yorker*

ONE EVENING IN November, 2021, a group of men assembled at sundown on the terrace of the Ruckomechi Camp, a safari resort on the Zambezi River. Since arriving by private plane, they had gone out lion-spotting, boated down the river, and landed a giant tiger fish; now they were clinking gin-and-tonics. Hippos wallowed in the water below.

The party was led by Renat Heuberger, a forty-four-year-old Swiss entrepreneur with narrow eyes and a cropped copper beard. Heuberger was the chief executive of South Pole, the world's largest carbon-offsetting firm, and he had come to Zimbabwe to fight off an urgent threat to his company.

A decade earlier, South Pole had signed a deal to sell carbon offsets from an effort to protect a vast swath of forest on the banks of Lake Kariba, upriver from the camp. The Kariba project, spanning an area ten times the size of New York City, was among the world's first "avoided deforestation" programs; by deterring local people from chopping down trees, it promised to prevent the release of tens of millions of tons of greenhouse gas. Leading corporations, including Volkswagen, Gucci, Nestlé, Porsche, and Delta Air Lines, paid South Pole nearly a hundred million dollars for Kariba credits, allowing them to market goods or services as "carbon neutral."

South Pole thus pioneered a model of carbon offsetting that has been counted among our best hopes for staving off climate catastrophe: a mechanism that diverts funds from polluters in wealthy countries to protect crucial ecosystems in the Global South.

Heuberger, a kinetic, grandiloquent man, speaks expansively about his mission. "We're here to save the climate," he told me.

As a child, Heuberger spent his spare time gluing protest flyers to car windows, and he considered himself an activist. But, as he built his company, he had developed a consumer-friendly brand of climate optimism. "It's not true that to save the climate we will all need to go into perpetual lockdown or stop having fun," he said, promoting Porsche's offsetting program. "In fact, it's the opposite"—drivers should enjoy their vehicles, knowing that "every ton of CO_2 they compensate for is backed by a verified emission reduction."

This perspective was enthusiastically received: Heuberger had regular speaking engagements at Davos and a spot in the World Economic Forum's network of experts. As brands scrambled for inexpensive ways to reduce emissions, the market for offsets surged, quadrupling in 2021 alone. That year, South Pole was approaching a billion-dollar valuation, which would make it the world's first "carbon unicorn."

But alarming news had reached the company's headquarters, in Zurich: it was at risk of losing its most lucrative project. By the terms of the Kariba deal, the company purchased carbon credits from a developer who oversaw the area's forestland, and sold them for a 25 percent commission. Now a competitor had offered the developer a substantial payment to take over the project. To help devise a response, Heuberger turned to an old friend from college, Dirk Muench, who had recently joined South Pole. Muench had left Wall Street to support climate action in the world's poorest places. He was a self-confessed stickler, and could be too fastidious for Heuberger's taste—but he was a skilled dealmaker.

When Muench heard the details, he was astonished that South Pole had done so little to secure its most important project. The entire agreement rested on a perfunctory contract that the developer, a white Zimbabwean tycoon named Steve Wentzel, could break anytime. To insure Wentzel's loyalty, Muench urged Heuberger to buy a stake in his business. They flew to Harare and took a chartered plane to the safari camp to conduct the negotiations in style.

Wentzel, a trim, chiseled man with a buzz cut silvering at the sides, was a former show jumper who had made a fortune in off-shore finance and then started investing in gold mines. On safari, he confided that he had no expertise in forest preservation; he had tried out carbon offsetting on a whim, when he was given a parcel of land as payment for a debt. (He told me the same story this July. "I don't know anything," he said. "I'm not a tree scientist or anything like that.")

Muench began to feel perturbed. A core principle of carbon offsetting holds that profits should be shared with local people, and South Pole maintains in its promotional literature that "communities living in the Kariba project area are the owners and main beneficiaries." But, as Wentzel described his indoor horse-riding range in Harare and his varied business interests, Muench wondered how much of the money that corporations spent on Kariba credits made it to the people on the ground.

As Muench probed Wentzel about the workings of his business, Heuberger was abashed. "If you want to survive as a businessman in Zimbabwe, you have to be a little bit of a special character," he later said. "We need to treat him with a little bit of respect." Wentzel told me that the inquiries didn't bother him: "I was, like, Yeah, whatever. As long as I get my end of the deal." Still, he had no intention of disclosing his financial practices. "You have your ways and means, and they're not all traceable, put it that way," he said. "No one actually has a damn clue about what's going on. Not even South Pole. I hold the key to Pandora's box."

As the men sipped their sundowners, Heuberger made Wentzel a striking offer. South Pole would pay about thirty million dollars for almost eight million credits. It would also open negotiations for a multimillion-dollar equity stake in his company, Carbon Green Investments. Wentzel agreed, and the mood on the terrace turned jubilant. Food and wine were ordered, and the celebrations continued by lantern light.

Muench acknowledged that, commercially speaking, the trip had been a "great success," but he couldn't shake a sense of unease about South Pole's work in Zimbabwe. "I realized, O.K., this is a huge money-making machine," he told me. Back in Zurich, he kept asking questions. "I said, 'Do you know what happens with the money?' And then someone told me, 'Dirk, you should look at the carbon side of this project—not just at the finances—to understand how bad it is.'"

The notion of carbon as a fungible commodity, like coffee or cotton, emerged in the late nineteen-eighties. As humanity reckoned with the harms of fossil fuels, a U.S. power company named Applied Energy Services conceived a novel way to reduce emissions: it could surround its main coal-fired power station with a forest, to absorb the carbon billowing from its chimney.

That plan turned out to be implausible. Scientists calculated that, to absorb the carbon the facility would pump out in its life span, the company needed to plant some fifty-two million trees—an impossibility in densely populated Connecticut. Then an executive named Sheryl Sturges had an inspiration: since the atmosphere was a global commons, why not situate the forest elsewhere? The company eventually paid for forty thousand farmers to plant trees in the mountains of Guatemala. It cost just two million dollars—pennies per ton of carbon.

Sturges's idea caught the world's attention. "Antidote for a Smokestack," a headline in *Time* magazine announced. A decade later, the concept of carbon offsetting was enshrined in international law, as thirty-seven industrialized nations and the European Union agreed to emissions-reduction targets under the Kyoto Protocol. Through the United Nations' Clean Development Mechanism, rich countries struggling to meet their goals could compensate by paying for projects in impoverished ones.

Growing up in Zurich, Heuberger had been terrified by the environmental calamities that defined the eighties and nineties: Chernobyl, the ozone hole, acid rain. He was a bright, sensitive boy who spent most of his time alone, biking in the mountains and memorizing train timetables. His anxiety about the threats to the planet became "paralyzing," he told me. When the Kyoto deal was signed, he had recently returned from a year as an exchange student in Indonesia, where he was "completely overwhelmed" by the experience of poverty. The prospect of a global trade in carbon struck him as a panacea—a way of using capitalist methods for radical aims. "Polluter pays, cleaner earns," he figured. "You could take the tools of the enemy and make them work for a better world."

Heuberger enrolled to study environmental science at the Swiss Federal Institute of Technology, where he found like-minded peers: young environmentalists who gathered to dream up gambits, from supplying the campus with organic coffee to publishing a magazine about sustainability. Among them was Muench, who was studying industrial engineering. "Renat was a little bit socially awkward, but also genuinely impressive," he told me. "He was like one of those activist environmentalists, whereas I was a typical business guy."

In 2002, as they neared the end of their studies, Heuberger, Muench, and a classmate named Patrick Bürgi were invited to a sustainability conference in Costa Rica. Preparing a presentation on carbon trading, Heuberger and Bürgi decided to make the concept more tangible by asking attendees to pay to offset the emissions from their flights. They got hold of a credit-card imprinter and ambushed delegates after each session: "Do you know that you emitted two tons of CO_2 by coming to this conference?" They raised more than ten thousand dollars. "It was quite easy to convince them," Bürgi said. "And then we started to get nervous—'What are we going to do with the money?'"

In the end, they donated the funds to the university hosting the event, to install solar heaters in place of a diesel boiler that supplied gym showers—an intervention that, by their reckoning, saved around seventy tons of carbon a year. "It was all kind of handmade and improvised," Bürgi said. "But it was so successful." Back home, he and Heuberger, along with a few friends, registered a nonprofit named MyClimate to continue offering offsets. (Muench, less confident in the project, left to pursue a career in investment banking.) They set up shop in the office of a supportive professor, launching a rudimentary Web site that allowed people to calculate their emissions and pay the appropriate penance. Carbon "tickets" were printed out and mailed to customers—until the venture became so successful that the professor complained they were using up all his toner.

By the time the Kyoto Protocol was ratified, in 2005, MyClimate was financing significant climate-action projects, including an initiative to supply clean electricity to a hundred Indian villages; it soon won a contract to compensate for emissions from the FIFA World Cup in Germany. The founders were still operating on a tiny budget, yet, as the UN's carbon-trading system got under way, they saw new possibilities. In a multibillion-dollar market, perhaps the profit motive would be the most effective way to spur action. "Capitalism works very efficiently," Heuberger said. "The idea that you could actually make money is a massive driver." The following year, he and Bürgi left their student venture behind and registered a company with three other friends, with a maxim of "profit for purpose."

The new business would focus on cultivating projects to sell credits through the UN system. The name, South Pole, referred both to Antarctica's melting ice caps and to the Global South, where most of its projects would be based. Heuberger borrowed twenty thousand francs from his parents for his stake in the business, and the entrepreneurs secured a workspace: a disused university chemistry lab, with long banks of sinks and cannisters of nitrogen that clients sometimes mistook for sequestered carbon. In the summer, they took breaks to swim in the Limmat, and in the winter they skied together.

Their first breakthrough came easily. China had just announced a five-year plan for renewable energy, and developers of wind, solar, and hydropower plants were gathering for the country's first Carbon Expo. Heuberger called a backpacking buddy who spoke Chinese and asked him to meet in Beijing. They printed business cards with a penguin logo, and scheduled meetings with developers in the lobbies of five-star hotels, though they were sleeping in a youth hostel.

South Pole's pitch was simple: it would help developers sell credits based on the carbon that would have been emitted if the power they produced had instead come from fossil fuels. The income would be small compared with what they made selling electricity, but it would come at no cost; South Pole would take care of all the complex carbon accounting, in exchange for a commission.

Heuberger's meetings at the conference led to several large projects, including a sprawling network of hydropower plants in the mountains of southwest China. After that, the company expanded rapidly. The founders scattered across Asia, Africa, and Latin America, signing up hundreds more projects. Soon, South Pole had opened branches in Thailand, Mexico, Indonesia, and India. Staffers, who came to be known as "penguins," greeted new employees with cries of "Welcome to the iceberg!"

In the years after the Kyoto targets came into effect, thousands of projects were registered under the UN's Clean Development Mechanism, and hundreds of millions of credits were issued, each worth one metric ton of carbon. Yet, as the market grew, so did questions about its integrity. Scholars worried that developers would inflate their projects' climate impact. Many environmentalists dismissed offsetting as a system of empty indulgences. One online spoof invited unfaithful spouses to pay someone else to remain faithful: "By paying Cheat Neutral, you're funding monogamy-boosting offset projects."

At parties in Zurich, South Pole's founders were grilled about the ethics of the carbon trade. "We were constantly challenged by friends," Bürgi told me. But Heuberger brushed aside such concerns. If humanity was to have any chance of saving itself, he was convinced, "there must be a positive narrative to climate action." When skeptics disagreed, he told me, his reaction was "Shut up. Keep it to yourself. Because we are on a mission here."

In 2009, South Pole attracted its first major investment, from BP. An executive in the oil company's alternative-energy division, Justin Adams, took a seat on South Pole's board, and made a close study of Heuberger. "Renat's a complex character, who I think is extremely strategic and thoughtful about how you can give people some hope in a time of fear," he told me. "But I suspect, like so many of us, there are deeper shadows in our own psyches. He had all the makings of a little emperor, and he's incredibly tough."

The association with a major oil company didn't trouble Heuberger. "We can talk to the biggest boys in the world," he said. With the market flourishing, he became increasingly focused on maximizing revenue. "I want to spin a big wheel," he told me. "With more money we can have more impact. We can do better, bigger things."

One day in 2010, an e-mail from Steve Wentzel, the Zimbabwean tycoon, arrived in the in-box of one of South Pole's founders—a tall, bluff German man named Christian Dannecker. It invited the company to enter new territory.

Wentzel ran a business in Guernsey that promised to "provide financial liberty through modern off shore financial services," as well as a money-lending enterprise based in Mauritius. He had recently acquired a parcel of woodland from a debtor who had failed to repay a large loan. The surrounding area was thronged with endangered wildlife, but the plot had been devastated by trophy hunting—"If you see a rabbit, it's a tourist," Wentzel said. At first, he thought that the land had little use. Then he heard about carbon credits and thought, "Let's see whether we can recoup our money that way." He found South Pole through Google.

Dannecker was thrilled. He had a passion for trees, and had often urged Heuberger to consider forest-carbon projects. "It's bloody tricky, but it needs to work, because otherwise climate finance will not reach those remote corners of the world," he said.

Dannecker flew to Zimbabwe with a team of experts to assess the possibilities. Wentzel's land was in the Binga district, south of Lake Kariba, an area threatened by economic turbulence. Zimbabwe was beset by hyperinflation—the South Pole team brought home a trillion-dollar bill as a souvenir—and by mass unemployment. Subsistence farmers were clearing patches of forest to plant crops, graze animals, and gather firewood. Dannecker and Wentzel figured that they could change those habits, largely by providing training in sustainable agriculture, and then sell credits based on the trees that they protected. "It's the bloody poorest area in the world that I have ever been in," Dannecker said. "This is where the money should go."

To secure cooperation in the area, flyers were distributed with cartoons of trees growing in the shape of dollar signs. Wentzel persuaded local chiefs to allow him to expand the project across four large districts, spanning two million acres of forest. In exchange, he promised that 70 percent of his revenues from the sale of carbon credits would be invested in Kariba and shared with the populace. "They all jumped on the bandwagon," Wentzel told me.

To Wentzel's mind, the people living in the forest would be getting money for nothing. "Don't cut the trees down, that's about the sum total of what they have to do," he said. "We don't ask them to get up in the morning, we don't ask them to do press-ups, we don't ask the birds to fly backwards. It is just a net positive for them."

He set up a company, Carbon Green Investments, to receive the sales proceeds from South Pole, and opened its accounts in the tax haven of Guernsey. "You have to use certain conduits," he told me. "Ultimately, my goal is to make sure that the project succeeds, and everyone gets the benefit. How it gets there? I'd rather you didn't ask those questions."

There is no hope of curbing the worst effects of climate change without saving our remaining forests. Earth's three trillion trees absorb nearly a third of humanity's carbon output, yet they continue to be destroyed at an alarming rate, releasing those stores back into the atmosphere. Forest-based offsetting rests on a simple

premise: if this carbon payload can be sold, it becomes more lucrative to leave trees standing than to cut them down.

Yet it is extraordinarily difficult to quantify how much carbon
these schemes really save. To do so, you must demonstrate that the
forest would have been razed without protection—a counterfactual
that is nearly impossible to prove. There are also issues of "leak-
age": even if the agents of deforestation are driven out of one area,
they may cut down trees someplace else. Then there is the question
of permanence. Greenhouse gases can linger in the atmosphere
for thousands of years—but forests are vulnerable to wildfires and
other calamities, and most protection schemes last no more than
a few decades. Twenty years after Applied Energy Services funded
the Guatemalan tree-planting project, researchers found that it
had largely failed. (AES disputes this.) The enormous amount of
land and labor devoted to forestry had led to food shortages, and
arguments had broken out; some farmers had simply refused to
plant the trees. In the end, the researchers calculated, the program had offset only about 10 percent of the emissions from the
coal plant in Connecticut.

The UN's carbon system allowed offsets in a variety of categories,
but it excluded forest-carbon projects, because of the particular
challenges of verifying their benefits. UN officials had deliberated
over an assessment framework called REDD—"reducing emissions from deforestation and forest degradation in developing
countries"—to distinguish the worthwhile forest-carbon projects
from the boondoggles. But, from the beginning, there was controversy over the science, and concern over the human cost of a forest-
carbon boom. White developers had begun buying up forestland
in the Global South, and reports emerged of "carbon cowboys" using violence and trickery to drive Indigenous people from their
territories.

Though the UN carbon-trading system never implemented the
REDD framework, it was taken up by a rival source of accreditation:
a nonprofit in Washington, D.C., launched by carbon-industry
players. The agency, which became known as Verra, had adopted
the UN's accounting methodologies, promising to apply them with
a lighter touch.

Verra allowed developers to choose among several different
ways to calculate the credits that their projects would generate.
That alarmed critics, who warned that developers would simply

select whatever model yielded the most credits. The agency's longtime chief executive, an environmental entrepreneur named David Antonioli, acknowledged the problem but told me, "If you require perfection, you'll have a hundred million dollars' worth of climate action. If you're more pragmatic about it, you might have two billion or five billion."

To register the Kariba project with Verra, South Pole had to predict how much of the forest would be lost without any intervention, and thus determine how much carbon the scheme would conserve over a thirty-year life span. Credits would be issued every year against that total, and the prediction would be checked once a decade, by comparing Kariba with an unguarded reference area nearby. South Pole's data analysts initially estimated that the program could save around fifty-two million tons of carbon. But Verra required them to rerun these calculations using one of its approved methodologies. The scientists used one named VM9, which generated a startlingly different projection: if the Kariba site was left undefended, deforestation would explode, resulting in the eventual loss of 96 percent of the forest. On that basis, the project would be eligible for almost two hundred million credits—four times the initial estimate.

Wentzel was delighted. At the time, the price of a single credit was about ten euros, suggesting that his cut might amount to hundreds of millions. The project began operating in 2011, opening a patchwork of community gardens and beehives across the site. "This is me on the way to getting this money back," he said. Then, with shocking abruptness, the carbon market collapsed.

"I remember watching the price curve in disbelief," Heuberger told me. "You'd check at 9 a.m., you'd check at 10 a.m., you'd check at 11 a.m., and every time it has fallen another five cents." By the end of 2012, the price of a single credit, which had peaked at twenty-five euros, had tumbled to thirty-nine cents.

Since the Kyoto Protocol had come into force, the market had been driven by government regulations, which obliged polluters that couldn't reduce their emissions to buy credits instead. When the financial crisis caused a slump in industrial activity, demand plunged.

Investors' confidence was further eroded by a series of scandals. Boiler-room scams selling fake offsets had sprung up across Europe, and hackers had penetrated the carbon registries of national governments to siphon off credits. One sprawling fraud, described by French police as "the heist of the century," had cost tax authorities five billion euros. After it was exposed, the Danish government admitted that 80 percent of the country's carbon-trading firms were fronts for the racket.

Even the legitimate programs inspired little confidence. The UN's Clean Development Mechanism had issued more than a billion carbon credits—three-quarters of which researchers later found to be environmentally dubious. Many of the projects were in China, including the sorts of renewable-energy schemes that South Pole was marketing. These plants, critics said, were too profitable to need carbon finance, and the credits they sold gave buyers license to go on polluting. Still more problematic were coolant factories that deliberately increased production of greenhouse gases, then profited by capturing and destroying them.

The greatest blow to the market was the failure of international climate agreements. The UN's Copenhagen summit, in 2009, was supposed to produce new binding emissions limits, but the negotiations collapsed. Three years later, the Kyoto Protocol's first commitment period skittered to a chaotic end, with almost half the participants having missed their targets and major players refusing to accept new ones. (The U.S. never ratified the agreement.)

"After ten years of believing that governments are lifting up the better world, the plug was out," Heuberger said. He resolved that the market must be rebuilt, but this time the private sector would have to take the lead. "The UN was gone, the governments were gone. We and Verra were still around, so we did it ourselves," he said. "Of course, it was not perfect. But it was the only show in town."

Heuberger summoned his team to a seaside resort in Krabi, on the coast of Thailand—a province of white-sand beaches, mangrove forests, and jungle islands. There, he outlined a plan to reposition South Pole. Rather than cater to clients struggling to meet government emissions caps, it would serve the so-called voluntary market—companies who chose to diminish their climate impact, for reasons of ethics or public relations.

Even with the market in retreat, South Pole planned to continue signing up new projects and stockpiling credits. Heuberger was convinced that the demand would soon rebound, and that his company would occupy a position of unchallenged dominance.

"Everyone else is dead," he said. "We are going to come out big here." As it happened, it would take years for the market to recover, but Heuberger was right that the absence of competition would provide an advantage. South Pole could "very easily contract with project developers, because there were no other buyers," Hannes Zimmermann, a former corporate-investment director at the company, told me.

As South Pole grew more dominant, some employees felt that it was straying from its purpose. Among the dozens of current and former staffers I spoke to, one quit after being asked to work with a chemical company whose carbon-offsetting proposals seemed to have no climate value. A second resigned after becoming concerned that South Pole was making exaggerated claims about its projects' benefits for local communities. Others objected to deals with commercial timber companies, which could earn credits by leaving tree plantations standing for a short time before cutting them down. One such scheme, which South Pole told clients would transform "degraded land in rural Mexico by sustainably growing teak trees," was run by a company that derives more than 97 percent of its revenues from logging.

Christoph Sutter, who served as the company's founding CEO before agreeing to share the role with Heuberger, had written a doctoral thesis on assessing the impact of offsetting projects. But, in six years leading South Pole, he had come to doubt the value of the carbon trade, particularly the sort of large renewable-energy projects South Pole was promoting. "I was building up this worry," he said. "It's just paper credits." He told me that he remained friends with Heuberger, but did not share his encompassing faith in offsetting. "The big majority of what you see in the market, in my view, boils down to a lot of greenwashing, a lot of marketing, a lot of money-making," he said. Heuberger had little patience for that sort of negativity. "Investors can smell it if the mood is not good," he said. Sutter quietly resigned in 2012.

In the depths of the crash, Wentzel became increasingly disillusioned. His income from credit sales had collapsed, and yet, by his estimate, it cost about sixty thousand dollars a month to pay for his project's staff and initiatives—community gardens, beehives, wells, fire protection. "I got myself into quite a big hole," he told me.

Local people were hardly lining up to thank him. In 2014, a community leader named Elmon Mudenda traveled to a Transparency International workshop in Harare and suggested that the Kariba project was a scam. "We have not seen anything really tangible," he said, according to the Zimbabwe *Herald*.

The following year, two Zimbabwean researchers traveled to the site to interview residents and published a damning study, "Struggles Over Carbon in the Zambezi Valley." They reported that the project's developers believed that "communal resources, including forests and wildlife, are there for the taking." The district councils, the study found, were "like a sleeping partner, with very little knowledge of what the project is all about and with no voice in its direction."

As pressure from the community mounted, Wentzel lost patience. One Saturday, he called South Pole and threatened to scrap the whole enterprise. "Carbon is nonsense," Dannecker recalled him saying. Wentzel demanded a million dollars by Monday to keep the project alive. Heuberger was at a friend's wedding in Italy when he heard the news. "We made a snap decision to ship the money," he told me. "Of course, we wanted something against that money. So we just took a few credits."

In fact, South Pole bought three million credits, at a low price of fifty cents apiece; it bought about the same number the following year. Ordinarily, the company made money by acting as an intermediary, selling credits on behalf of developers and charging a commission. Now it bought the credits directly, meaning that it would keep all the profits when they were sold. Wentzel told me that, after receiving the funds, he drove through the Kariba site, settling his obligations with bundles of cash. "A hundred thousand dollars is only as big as a brick," he said. "It's not difficult to carry it around."

The intervention revived the project. Mudenda, the community leader who had previously criticized Kariba, now praised its beekeeping efforts to a local newspaper; villagers were making four hundred dollars for harvests of organic honey. In 2016, Dannecker flew to Zimbabwe, and blogged about his visit under the heading "Why I Get Out of Bed Every Morning." The area had undergone a drought, but the project's wells had provided water, and its farming efforts had yielded food. "Sometimes, the business feels like what it is—a business," he wrote. "But it's actually much more: our business has a purpose."

Yet the company's business in Zimbabwe rested on a shaky foundation. Project monitors had surveyed the site and found fewer trees than South Pole planned to claim credit for. Then Dannecker learned something even more alarming: the rate of forest loss in the project's reference region—the benchmark against which its success would be measured—was starkly lower than projected. The wave of deforestation that South Pole's efforts were supposed to prevent was looking more like a trickle, which could significantly diminish the value of the project.

Dannecker decided not to do anything rash. "There was no urgency, for two reasons," he told me. First, there were still years to go before South Pole was due to check its model against reality. Second, amid the market slump, "there was no bloody demand" for the Kariba credits anyway. But that was about to change.

One Friday in August, 2018, a small figure in a yellow raincoat shuffled up to the Swedish Parliament in Stockholm and sat down at the foot of the building. Beside her was a sign painted with block letters: "SKOLSTREJK FÖR KLIMATET."

Greta Thunberg's "school strike" represented something rare in an age of futility: an individual act that reverberated around the world. The movement that she inspired, in which millions of children skipped class to demand climate action, gave rise to the biggest environmental protests in history. To Heuberger, himself a former child activist, the fifteen-year-old Swede seemed like a kindred spirit. "Greta Thunberg and the climate strike were of outstanding importance," he said at the time. "Today, no listed company can afford to be on the sidelines when it comes to climate protection."

The strike, playing out amid wildfires and increasingly apocalyptic weather, put renewed pressure on the world's largest companies. South Pole's business soared. "Going carbon neutral is the latest luxury trend," the company proclaimed, after Gucci announced that it would use South Pole credits to cancel out the emissions of its supply chain. Nestlé soon declared that Kit Kats and Nespresso pods would become carbon neutral. Porsche assured customers that the emissions from ten thousand miles in a Cayenne could be scrubbed for as little as sixty-seven dollars. "Climate action has to move away from this idea that only completely green people are part of it, the kind of people wearing hemp shirts and walking everywhere," Heuberger said, in an interview promoting the scheme.

Even after Thunberg denounced offsetting as "a dangerous climate lie," her movement fueled the market. JetBlue announced that it would use South Pole credits to help offset the emissions from its U.S. flights. Delta followed with a billion-dollar pledge. Scores of other companies used South Pole offsets toward their net-neutrality claims.

Though South Pole has a portfolio of more than a thousand projects, its partnership with Wentzel proved singularly lucrative—not least because of the credits it had bought directly during the slump, for fifty cents apiece. The sale of these credits provided no additional funding to the project, but the margins for South Pole were huge. Kariba credits would ultimately be worth more than fifteen dollars.

Heuberger liked to say that business was like surfing: you waited for the wave to come, and then you rode it all the way. South Pole launched a major expansion, growing to twelve hundred employees and twenty-nine international offices. To celebrate the opening of its New York branch, it announced that it had made the entire city carbon neutral for one hour. It repeated the trick during the city's 2019 Climate Week—but this time it claimed that, for the span of one second, it had neutralized the emissions of the entire world.

The carbon market grew sevenfold after the school strike, to two billion dollars a year—but that was still smaller than global sales of nail clippers, or fireworks, or pepper. In September, 2020, a new initiative emerged that promised to radically expand the trade.

The Taskforce on Scaling Voluntary Carbon Markets was a corporate powerhouse, led by Mark Carney, the former head of the Bank of England. It had some four hundred members, including many of the world's largest suppliers of fossil fuels. Oil companies presented particularly voracious demand; their net-zero pledges required cancelling out several billion tons of carbon a year—far more than the world's supply of offsets. Carbon credits had typically been sold from a single developer to a single buyer, often with the aid of an intermediary like South Pole. Carney's vision was to effectively create a stock market for offsets, so that they could be traded with the same speed and ease as any other financial instrument.

By the logic of the markets, such trading would help facilitate funding to environmental projects. In practice, it often diverted funds to speculators. BP and Shell had opened carbon-trading desks, and the Saudi government did, too. Gilles Dufrasne, of the nonprofit Carbon Market Watch, observed that credits could be traded over and over before being used to offset emissions: "When you buy a carbon credit, what is the chance that somewhere in the value chain it was once owned by Shell, and that some of what you pay represents the cut they took?"

When the task force held a promotional event at a UN climate summit in Glasgow, Thunberg and other protesters were filmed outside, singing, "You can shove your climate crisis up your arse." Later, she tweeted an addendum: "I've decided to go net-zero on swear words and bad language. In the event that I should say something inappropriate I pledge to compensate that by saying something nice."

Despite the controversy, South Pole was quick to capitalize on the demand. It made deals with Gazprom and Chevron. TotalEnergies announced that it had delivered its first shipment of "carbon neutral liquefied natural gas" using Kariba credits, and the Dutch provider Greenchoice bought millions more to market its gas as "sustainable." South Pole's willingness to do business with energy giants rankled its workforce. The staffers were largely young and idealistic, and many felt that the company was helping the world's worst polluters repair their image. Heuberger defended the choice: "Why wouldn't those guys who make gazillions of dollars of profits put a portion of that money into funding climate action?"

The head of South Pole's consultancy division, Rebecca Self, was concerned that the company was awarding "Climate Neutral" badges to clients who seemed to be making little meaningful effort to cut their emissions. But, she told me, when she raised these objections, Heuberger accused her of "sounding like an N.G.O." and "trying to kill the projects." (Heuberger maintains that he does not recall saying this, but in our conversations he repeatedly condemned environmental nonprofits, complaining about their "destructive all-out bashing" of the carbon market and even suggesting that such organizations are secretly chaos agents funded by the oil industry.) Not long afterward, Self learned that South Pole had helped the Qatar World Cup substantiate a carbon-neutrality claim that excluded most of the emissions from the construction of seven air-conditioned stadiums. She put her concerns in writing and resigned.

South Pole's growth continued unimpeded, and it acquired five smaller rivals. The founders began selling their own shares to wealthy investors, including the government of Singapore, the Liechtenstein royal family, and Salesforce. A deal with Swisscom secured South Pole's billion-dollar valuation, though Heuberger told me that he came to regret the status it conferred. "As long as we were a startup, we were everybody's darling," he said. "People think if you're a unicorn you must have made a shitload of money and completely ripped everybody off." Still, he remained afraid of being outpaced by some fast-growing competitor. "We have to defend our market share," he reasoned. "Doubling is for losers, because everybody else is tripling."

When Dirk Muench went to work at South Pole, in May, 2021, it felt like a homecoming. In his early days on Wall Street, he had been enthralled by the "prestige and grandeur," he told me, but he had come to see all that as "smoke and mirrors." He had left JPMorgan, studied climate science at Columbia, and eventually reached out to his old roommate Patrick Bürgi. "What I really want to do is try to direct large amounts of capital towards climate projects," Muench said.

Hired as the head of corporate investments, he rejoined his college friends at the Technopark, a hulking research complex in Zurich that had housed South Pole's first office. The company now occupied a much grander space there, but the founders still swam together in the river. "Everything felt really good," Muench told me.

Gradually, though, he realized that South Pole was far less deeply involved than he had believed in the projects it presented as its own. "They have created this image of being a project developer that protects the climate," he said. "They are nothing of that sort. They are a broker." Heuberger was hard to recognize as the radical environmentalist he had admired in college. "I started to see that he's lost," Muench told me. "If you're in a business that's so lucrative, so good, and you've gained your power, your status, and your money from that, you do all you can to protect it."

When Muench went to Zimbabwe to salvage the Kariba project, he returned even more troubled. Asked to perform due diligence on the investment South Pole planned to make in Wentzel's company, he pulled the records of payments to Kariba and saw that all the money—some forty million dollars—had been wired to a single account in Guernsey. He told Wentzel that he needed evidence of where the funds from that account had gone. But, after six months of e-mails and phone calls and another meeting in Zimbabwe, Wentzel remained evasive. South Pole had hardly any idea what had happened to tens of millions of dollars its clients had spent supposedly offsetting their carbon emissions.

The published project literature outlined what should have happened to that money: Wentzel's company was to keep 30 percent, and the rest was to be used to pay district councils, fund project activities, and top up a rainy-day fund. Some of it had clearly been spent on the Kariba site. Along with the farming activities, there were new school huts and clinics, anti-poaching patrols, and fire-suppression measures. But, Muench said, when Wentzel finally sent a spreadsheet of his spending, in the summer of 2022, it accounted for only around six million euros. Even that, he told me, had "no backing, nothing behind it."

On July 9, Muench sent an e-mail to Heuberger and other executives, with the subject line "Red Flag." He reported that, after a long investigation, he could only conclude that most of the funds for the Kariba project had gone astray. In private, he says, he urged Heuberger to admit the problems: "We need to come public before it is in the press." Heuberger was disinclined to listen to Muench. "He's very driven—the mindset is totally on impact," Heuberger told me. "For him, there's only good and bad." South Pole removed Muench from the inquiry into Wentzel's finances, after which "the relationship and flow of information proved to be more productive again," the company said. To Heuberger's mind, the Kariba project was as good as it needed to be: "Is it perfect? Is the guy a hundred percent? Every dollar always a hundred percent?" He shrugged. "You have to navigate your way around."

In September, Wentzel flew to London on business, and I met him for breakfast at a bistro on Sloane Square. He wore designer chinos, Chelsea boots, and a crisp white shirt, but when he greeted me I noticed that he was missing a front tooth.

I had spent hours in two previous encounters questioning Wentzel over the project's finances. At first, he had reeled off contradictory figures, getting out a calculator and punching the keys before giving up and pushing it away. In the end, he admitted that his inability to account for the money was no accident. "There's no paper trail," he told me.

Years of political and economic instability had made it too precarious to bank in Zimbabwe, he said, and transferring money to a sanctioned state from Guernsey was a bureaucratic headache: "Do you know how much compliance I had to go through to just have one transaction?" So he had devised an untraceable way of moving the funds. "It was illegal," he acknowledged, "but it got looked over."

When he needed money for the project, he said, he would transfer it from Guernsey into the account of an acquaintance who wanted electronic funds, in "Mauritius or the Cayman Islands or the Seychelles or Russia or wherever," and they would arrange for the equivalent in U.S. dollars to be delivered to him in Zimbabwe. Other times, he would pay an invoice for someone else—for a consignment of motorbikes, perhaps—and that person would deliver him the same amount in cash.

"This looks really bad, because you're just sending money here, there, and everywhere, but, on the receiving side, I can show where we've got it," he told me. "Well, I can show you the bundles of cash on the floor." When payments arrived, he said, he would "grab the money and run with it," distributing it among the stakeholders in the project. "For any kind of European or American, that's not comprehensible," he said. "How many Western people have carried half a million dollars of cash in their hand?"

Wentzel's demeanor seemed to lighten as he unburdened himself, and he began to stage a mock interrogation. "Can I see the SWIFT?" he boomed, referring to the code that banks use for international payments. "The money got there swiftly, but I can't tell you what the SWIFT was." Suddenly his waggish smile gave way to a frown. "I don't know what you're going to report on this, and I hope to God it's not all of it, because I probably will go to jail," he said. Then he reassured himself. "I'll go to jail for the right reasons," he said. "Savior or villain? I'm right in the damned middle. And I'm happy to be that way."

The Castello di Modanella is a twelfth-century castle whose battlements rise from the Tuscan hills. It claims to have hosted at least two Popes, and Galileo is said to have been banished to its tower after being convicted of heresy. Its stone chambers are now mainly used for gala dinners and, in recent years, for South Pole's management retreats.

On a warm night in September, 2022, as staffers partied in the gardens, a circle of weary executives, including Heuberger, Dannecker, and Muench, stood in a courtyard. They were reckoning with bad news.

The previous year had marked a decade since Kariba was launched, which meant that South Pole was required by Verra to check its explosive predictions against reality. After months of reviewing satellite imagery, the company's data analysts had determined that deforestation in the control zone was dramatically lower than projected. They estimated that only fifteen million of the forty-two million carbon credits generated by the project had actually been backed by avoided emissions. All the rest of those supposedly offset tons of carbon simply weren't real.

Muench and another executive urged Heuberger to stop selling offsets from the Kariba project immediately. "If it comes out that we've knowingly sold credits that weren't equivalent to a ton of CO_2 emissions avoided, it would do huge damage," Muench said. Heuberger rejected that idea. The credits had been validated by Verra, he argued: "If you want to scale, you have to rely on certain rules and systems." (South Pole acknowledges that this conversation occurred but says that it took place after the trip to Tuscany.) Back in Zurich, South Pole continued enthusiastically promoting Kariba. In the months after learning of the miscalculation, it sold more than three million environmentally worthless credits, to Porsche, Nestlé, and Nando's, along with others including the Cannes Film Festival and a network of Australian zoos.

But scrutiny of the market was increasing. Investigations by the *Guardian,* Bloomberg, and others had highlighted questionable accounting and community abuses by carbon projects. Greenpeace and other nonprofits had published reports denouncing the trade as a dangerous distraction from genuine efforts to reduce reliance on fossil fuels.

One Friday evening that November, a forest-ecology expert named Elias Ayrey posted a satellite image of the Kariba area online. "I find myself quite upset," he wrote. "I just reviewed a #carbon project that's likely receiving more than 30x as many credits as it should." Ayrey, who works for an independent ratings agency called Renoster, had used NASA satellite imagery to calculate that deforestation in the Kariba reference region was significantly lower than the company had stated. He signed off with a disclaimer: "All opinions are my own. And my own opinion is that everyone involved with this project should be arrested."

Late that night, Heuberger posted a caustic response: "It looks like you really don't understand how carbon finance works, and your only goal is to criticize and spread fake news." His phone soon started buzzing with calls from alarmed clients, and four days later the company finally instructed staff to pause the sale of Kariba credits. By then, South Pole had off-loaded twenty-three million credits from the project—eight million more than it could justify.

The mood around the Technopark was tense. One day that fall, Muench sat down next to Heuberger in the open-plan office and said that he was still worried about Kariba. "You sold credits that weren't real," he said. "They didn't have an impact on the climate that you expected, and, on top, you guys took a lot of money."

Heuberger was enraged, he told me: "He came and said, 'Renat, you enriched yourself. I demand that you come out and rectify all those omissions transparently. Talk to your investors and your clients and hand them back their money.'"

Muench said that Heuberger screamed at him, "Get out! Get out! You know nothing!" But then he followed his old friend into the street and persuaded him not to leave.

In December, South Pole organized an all-hands meeting to quell the staff's concerns about Kariba—to "build trust around this truly amazing project," as one executive said. (A recording of the session was shared with me by Follow the Money, an investigative newsroom in the Netherlands.) Christian Dannecker began his talk with a convoluted disquisition on deforestation curves, counterfactual modelling, and the limitations of NASA satellite data in assessing dryland deforestation. Then he took questions.

"Are the Kariba credits based on reality?" one staffer asked.

"I give the question back: What is reality?" Dannecker snapped.

South Pole's head of U.S. sales tried again: "We have clients coming to us saying, 'Hey, the credits that you sold us, that impact that we claimed, did that actually happen? Yes or no?'"

A public-affairs executive intervened: "Which is why we're holding the current verification at the moment—because we want to make sure that that is the case."

"How much profit has South Pole made by selling Kariba credits?" another staffer asked.

"I didn't do the numbers, to be honest," Dannecker replied. "I guess we probably made ten million dollars." There were audible gasps, before another executive warned, "To be clear, we don't want to repeat that publicly."

Muench watched the meeting in disbelief. "It was Machiavellian," he said. "I think, at the end of the day, they started to lie to themselves." He left the company three days later, after filing a report through its whistle-blower channel. He received a brief response a few weeks later. "An investigation has been completed and we conclude that South Pole were following the approved Verra methodology," it read. "We are therefore going to close this case." That month, Verra certified another seven million credits for the Kariba project.

In the absence of a government regulator, Verra had become the primary standard-setter of the voluntary carbon trade. The agency controlled about two-thirds of the market, and nearly half its projects were forest-carbon schemes like Kariba. As the global trade ignited, the scenes inside its head office had been chaotic. "One day we were just bumbling along, and then the next day it was drinking from a fire hose," Andrew Beauchamp, who worked at Verra for eight years, told me. The chief executive, David Antonioli, a rangy man with an elastic grin, said that developers applied increasing pressure to wave their applications through. "They got pretty worked up," he said.

For years, Verra had delegated the oversight of projects to outside environmental auditors. (The Kariba project had been reviewed five times, which Heuberger often cited as evidence of careful supervision.) But the auditors were hired and paid by project developers, potentially creating an incentive to ignore concerns. When Verra staff ran "sniff tests" on some of the auditors' reports, they found scores of serious errors. "There are a lot of us who believe that this resembles an elaborate fraud," Danny Cullenward, a climate economist at the University of Pennsylvania, told me. Verra allowed credit sellers to claim that they had undergone a robust certification process, "even though every element of that process, when you dig in, is conducted by financially self-interested parties."

Many observers felt that Verra had an even greater conflict of its own: it charged a fee to certify each credit. "The more credits they issue, the more money they make," Niklas Kaskeala, a founder of Compensate, a Finnish nonprofit focused on carbon-market integrity, told me. "It's structural corruption." (Verra recently began overseeing auditors more thoroughly, and is working to revise its procedures around forest carbon—but only after verifying more than a billion credits.)

This January, the *Guardian* ran a story that alleged, on the basis of three scientific studies, that more than 90 percent of the forest credits Verra had certified were "worthless." Though Antonioli resigned soon afterward, Verra disputed the report, arguing that most forests protected through its programs are still standing, despite commercial pressures to cut them down. Several independent agencies and academic studies have since concurred that most forest-carbon projects in existence are selling offsets based on vastly inflated claims. By those reckonings, several hundred million tons of carbon that were supposedly offset will linger in the atmosphere for centuries.

As 2023 began, South Pole was struggling to control the damage to its reputation. It had invited clients to visit the Kariba site, and Dannecker had published a blog post presenting the project as a resounding success. Then, in January, a tape of the staff meeting about the project was leaked to Follow the Money. The resulting story ran under the headline "Showcase project by the world's biggest carbon trader actually resulted in more carbon emissions."

South Pole responded with a lengthy rebuttal, complaining of "exaggerated and misleading reporting." Yet its own statements about its profits were tangled. Dannecker had written in his earlier blog post that the company took a 25 percent commission on the Kariba-credit sales and sent the remaining forty million euros to Wentzel, for distribution. Now he acknowledged that the company had made far higher margins on the credits it had bought directly, claiming Wentzel had received fifty-seven million euros, from total revenues of more than a hundred million.

Within South Pole, the handling of the Kariba project inspired dismay, and several staffers resigned. The public, too, was increasingly suspicious of the voluntary market. Consumer groups began suing companies for greenwashing, and the European Parliament proposed banning claims of net neutrality based solely on off-setting, after finding that 40 percent of green marketing in the E.U. was "completely unsubstantiated." Gucci quietly dropped its carbon-neutral claim. Volkswagen, Barclays, L'Oréal, and McKinsey said that they would stop buying offsets from the Kariba project. As the price of carbon credits tumbled, Heuberger was aghast. "This is now a loose cannon—it's spinning, spinning, spinning," he said. "We have to stop the feed."

A few weeks after the details of the all-hands meeting came out, Muench received an e-mail from South Pole's lawyers, demanding that he present himself for questioning about the leaked tape. He denied being behind the leak, but the investigation struck him as an opportunity to relay his concerns about Kariba. He agreed to meet the lawyers, and sent them detailed written testimony in advance. The lawyers abruptly cancelled the meeting.

Muench said he heard nothing more about the accusations against him. But, when I met Heuberger this summer, he told me that the Kariba controversy had been concocted by his former friend. "It's very psychological," he said. "To his frustration, we were more successful than him. And in his mind there was this story created—like, the only reason South Pole is successful is because it's a weird company. It's a little bit fishy. We are cheating." (I first spoke with Muench after Heuberger mentioned the concerns that he'd raised, but he declined to speak publicly. He eventually agreed to do so only after Heuberger continued criticizing him.)

South Pole insists that all the Kariba credits it sold will ultimately be backed by a real emissions reduction: Verra's methodology allows it to pay back the over-issuance with offsets generated in the future. But several experts, including sources at Verra, told me that this may be impossible, since the drop in deforestation in the reference region will severely limit the project's eligibility for new credits.

Justin Adams, the former BP executive who had sat on South Pole's board, told me after first hearing about the controversy that he believed the company was being unfairly maligned. "Have they got all the calls on Kariba right? No. But have they been a net positive force in a world that's full of darkness right now? Of course they bloody have," he said. Yet as he learned more he began to lose sympathy. "As a billion Euro company (a fabled unicorn) SP should have had far greater control and audit mechanisms," he wrote to me. "Certainly they took a lot of risk early on but the rewards in later years look out of whack. More commensurate with an oil and gas concession than an environmental and community development project."

People in the area suggest that the project's impact has been minimal. Iain Foulds, a white Zimbabwean farmer who has become an outspoken critic of Kariba, told me, "It's nothing to do with the environment or safeguarding our flora and fauna. This was about money. Yes, there's a bit of petty cash filtering in here and there for their little projects. But the big bucks—where's all that going?" Nevertheless, community leaders are wary of losing whatever benefit they receive. In June, I spoke with Elmon Mudenda, who joined a video call from a dimly lit room, with cracked plaster and peeling paint. His great fear was that the project would be deserted. "At the end of the day, who suffers is the community. Nobody else," he told me.

The next month, Follow the Money published another story. Reporters had visited the project site, and disaffected locals had shown them a series of abandoned vegetable gardens. The lead author, Ties Gijzel, told me he and his reporting partners had also heard that Wentzel was involved in trophy hunting.

Big-game hunting is legal in Zimbabwe, but South Pole has portrayed the Kariba project as a haven for wildlife, protecting "numerous endangered species such as the African elephant, the lion, the hippopotamus." When I questioned Wentzel, he acknowledged that trophy hunting occurs across the project area. He had taken control of the sport himself in one region, granting rights to an operator named Dalton & York, whose Instagram page has dozens of images of hunters displaying dead lions, elephants, and crocodiles. In one, a grinning man holds up the carcass of a leopard as blood drips down his forearms. A video shows hippos being shot through the head as they wallow in the Zambezi.

Over the summer, the Zimbabwean government announced plans to centralize control of carbon-trading projects within its borders, and to seize 30 percent of future profits. Wentzel told me that he didn't much care what happened to Kariba. "It's no skin off my nose," he said. He was already planning a new venture, which he described as "Kariba on steroids." In his scheme, local women might be employed to stitch car mats for Porsche or to sew beads onto Gucci accessories—though he confessed that neither brand had agreed to the endeavor. He was "struggling to think of something for Nespresso," he said, but he had been working on a deal to sell driftwood lampshades to an American furniture catalogue. He has already registered the new business in Ireland, under the name Fair Share.

Sylvera, the most prominent of a small group of ratings agencies that are seeking to improve transparency in the carbon market, occupies a plant-filled office in an unassuming building in London. When I visited this June, I was met by the chief executive, Allister Furey, a round-cheeked man in sneakers and a vaguely psychedelic T-shirt.

Furey, a neurobiologist with a Ph.D. in machine learning, worked for a decade in renewable energy before becoming fixated on carbon removal. In order to limit global warming to 1.5 degrees, the UN has said, humanity will need to find a way to suck around ten gigatons of carbon a year out of the atmosphere. "The scale of the challenge is so extreme," Furey told me. "You need to move trillions and trillions of dollars." Carbon offsetting seemed like a viable source of that kind of funding, so Sylvera was founded in 2020 to stimulate investment by sorting out the good credits from the junk. Since then, using spaceborne radar and satellite imagery, it has estimated that some 80 percent of the forest-carbon projects it rated were likely over-crediting. "People get paid more for issuing a higher number of credits without going to prison, so there's a very strong incentive," Furey said.

Recently, two new initiatives have arisen to encourage higher standards. The Voluntary Carbon Markets Integrity Initiative, launched with backing from the British government, set guidelines to deter corporations from making vacuous claims about the benefits of offsetting. The Integrity Council for the Voluntary Carbon Market, a successor of Mark Carney's task force, announced a set of "Core Carbon Principles" to assess the quality of existing schemes. "If you build integrity, scale will follow," Annette Nazareth, the council's chair, told me. But an analysis by the carbon-data firm Trove Research found that 95 percent of projects on the market would fail to meet these new standards.

Yet offsetting remains central to global plans to reach net zero. The Paris Agreement, signed in 2016, envisaged a new carbon-trading framework—though the details are still being debated seven years later. More than two-thirds of participating countries plan to use offsets to meet their goals, and an alliance of governments and industry figures is lobbying for forest-based projects to be included in the new system.

Experts told me they feared that the same registries they blamed for catastrophic mismanagement of the voluntary market would, by virtue of sheer convenience, end up taking a central role in the UN process. Axel Michaelowa, a researcher at the University of Zurich who co-authored several UN climate reports, said Verra had been telling officials that, if they used its verification services, "you'll have a one-stop shop—you don't have to pay a single cent." That, he said, "would of course mean that all the shortcomings of these private programs would contaminate the compliance market."

This November, when world leaders gather in Dubai for the UN's annual climate summit, they will seek agreement on the principles of the new system. To lead the summit, the host country, one of the world's largest exporters of fossil fuels, has appointed the head of the Abu Dhabi National Oil Company.

One afternoon this July, I met Heuberger at the Technopark. The sky over the city was cobalt blue, and the air felt strikingly clean. We strolled along the Limmat, passing city workers shedding their suits to dive into the water, then turned up a set of dilapidated steps to his house—an unobtrusive cream-colored structure, set into the hillside above the river. When I remarked on its modesty, he nodded. "It was not cheap, but it's way far away from the villas you can see," he said, gesturing at the mansions up the hill. Heuberger also has a pied-à-terre near Davos, but he insisted that the allure of business had never really been about money. He saw building his company as a kind of game: "All of us are driven, of course, by winning."

For the moment, Heuberger was winning his latest fight—what he called the "shitstorm" in the carbon market. He blamed a handful of enemies: Muench, environmental campaigners, oil companies funding secret plots to destroy those seeking to put a price on carbon. As he enumerated these annoyances, he batted his hands in the air, as if swatting gnats. "We are stubborn, whack-a-mole people," he said. "You can slap us as many times as you want. We always come up." In recent weeks, Heuberger had been working to "change the narrative." He had revealed plans to collaborate on carbon-removal projects with Mitsubishi, and had traveled to London for Climate Week, where he joined a panel on preventing misinformation around offsetting. Though his company had spent years awarding its clients badges that proclaimed them "carbon neutral," Heuberger dismissed this designation onstage as "an easy catchphrase." The times demanded a new approach, he said: "We have to become robust and honest." To that end, South Pole was launching a new insignia. Its penguin logo would now be encircled with the words "This Company Funds Climate Action." The announcement was met with cautious applause.

At his house, Heuberger led me through a garden overgrown with lavender and brambles, and into a sparsely furnished living room, where a baby's play mat was the only splash of color. He has four daughters with his wife, Zani, whom he met in Indonesia while scouting projects for South Pole. Their youngest was born last year. When he talked to his daughters about climate change, he said, he focused exclusively on messages of empowerment and optimism: "There's no climate anxiety at all in our house."

On a deck with sweeping views of the city, Zani was grilling sausages, halloumi, and tempeh kebabs. After dinner, Heuberger sat sipping wine with his feet tucked under him, and grew a little maudlin as he gazed out over the dimming skyline. "We were eating dry bread for many years, happy that we just survived," he said. "I'm sounding like an old guy now. But I look at those young kids who come with their blockchain-enabled super-transparent solution, and say everything has been shit, what we did. O.K. Calculate Kariba credits. Good luck. It's not that easy." In my earlier conversations with Heuberger, he had parried every perceived slight or criticism, but as we kept talking those defenses seemed to slip. He confessed that it had been hard to maintain his positivity. "When you think you have done the right thing, something goes wrong, and somebody says you had bad intentions—that's the most harmful," he said. "You have to build a thick wall between you and your true feelings, which are of course depressing."

He told me he was spending a lot of time cycling these days, on the same battered Titan road bike he had bought with saved-up pocket money as a boy. "I had my beliefs and my convictions and my bicycle," he recalled fondly. "I was in my own world." To lift his spirits, he had also taken up the cello again—another childhood pursuit—and started singing lessons. "Except for me, everyone else is a little kid at the class," he said. In a few weeks, he was due to appear in an amateur opera, a production of *Hansel and Gretel.* Heuberger had been cast as the children's father—"the bad guy, the weird guy"—who returns triumphantly from a profitable day at the market to find that his children have been abandoned alone in the forest. "He doesn't really understand," Heuberger said. "He thinks he's doing a good thing. But he doesn't get it at all."

CAROLYN KORMANN

Why Maui Burned

FROM *The New Yorker*

AT 4 P.M. on August 8, Shaun Saribay's family begged him to get in their car and leave the town of Lahaina, on the Hawaiian island of Maui. The wind was howling, and large clouds of smoke were approaching from the dry hills above the neighborhood. But Saribay—a tattooist, a contractor, and a landlord, who goes by the nickname Buge—told his family that he was staying to guard their house, which had been in the family for generations. "This thing just gonna pass that way, downwind," Saribay said. At 4:05 p.m., one of his daughters texted from the car, "Daddy please be safe."

Within ten minutes, it became clear that the fire had not passed downwind. Instead, towering flames were galloping toward Saribay's house. He got in his truck and drove to Front Street—Lahaina's historic waterfront drag—and found gridlock traffic. Saribay, a stocky forty-two-year-old man with a tattoo covering the left side of his face, texted his daughters. "Don't worry. Dad's coming," he wrote. Then he lost cell service. At 4:41 p.m., he pulled into the one large open space he could find, a parking lot behind the Lahaina United Methodist Church, which had just started to burn.

Saribay had recently built a closet at the church, so he knew where all the water spigots were. He filled buckets and water bottles and scrambled to find neighbors' garden hoses. With the help of three other men who had retreated to the lot, he soaked the church, again and again, fighting a three-story ball of fire with the equivalent of a water gun. At times, the men were stomping, even peeing, on sparking debris. Saribay recorded a video for his kids:

"It's bad. All around—crazy," he said, panning the hellscape behind him. "Remember what Dad said, eh? I'll come back." Almost as if to reassure himself, he added, softly, "I know you guys safe."

Saribay recorded videos throughout the night as he fought the fire. Despite his efforts, flames consumed the church. Well after midnight, the men tried to save a neighboring preschool, but that caught fire, too. When the sun rose and the wind began to ebb, Saribay got on an old bike and rode around town looking for other survivors. "I'm seeing fucking bodies every fucking way," he recalled. "I'm pedaling through charcoal bodies and bodies that didn't have one speck of burn—they just died from inhalation of black smoke. I felt like I was the only fucking human on earth."

The wildfire in Lahaina was the deadliest in the United States in more than a century. Ninety-nine people have been confirmed deceased, although for weeks the death toll was thought to be even higher, with police reporting that more than a hundred bodies had been recovered. In a town of nearly thirteen thousand people, at least seventy-two hundred were displaced. Twenty-two hundred structures were damaged or destroyed, and the estimated cost to rebuild is five and a half billion dollars. "I have been to most major disasters in the United States in the past decade. This is unprecedented," Brad Kieserman, a senior official with the American Red Cross, said. "The speed of the fire, the level of fatality and physical destruction, the level of trauma to those who survived—it's unspeakable."

The destruction may have been unprecedented, but the fire itself was not. Public-safety officials, scientists, and activists had warned for years of the wildfire risks in Maui, owing to the growing population and the dryness of the island. "It was a ticking time bomb," Willy Carter, a conservationist who studies native Hawaiian ecosystems, said. "The bomb went off." Weeks before the disaster, conditions in parts of the state had been categorized as "severe drought," and on August 4 the National Weather Service warned of hazardous fire conditions in the coming days. With a high-pressure system north of Hawaii and Hurricane Dora spinning hundreds of miles to the south, forecasters predicted that strong winds would be blowing, allowing flames to spread fast.

At 12:22 a.m. on August 8, a brush fire ignited in Olinda, in the mountains of Central Maui, prompting evacuations. At 6:37 a.m., thirty-six miles away, another brush fire ignited, in a bone-dry field bordering Lahaina Intermediate School. Hard winds had toppled utility poles, and flying sparks from downed power lines likely started the blaze. (The official cause is still under investigation, according to the Maui Fire Department.) Nearby residents were ordered to evacuate within three minutes. By 10 a.m., the county announced, via Facebook, that the Lahaina fire was "100% contained," but that a main road was closed.

Around 3 p.m., people noticed smoke clouding the sky near the school. With the wind gusting more than seventy miles per hour, the fire had flared up again in the same area. During the next hour, the fire hit "crossover"—a term used to describe a moment when the relative humidity drops below the temperature in Celsius. This allowed the blaze to tumble freely and grow exponentially faster, exceeding firefighters' capabilities. All they could do was try to save lives.

It would be difficult to overstate the horror of these hours, the disorientation of the hazy twilight caused by toxic smoke, the searing wind and glowing ash, the stark terror of being surrounded by tall flames, the suffocation. At various times, Maui police, in coordination with the power company, closed most of the roads out of town, because of tangles of downed lines and branches, but also because of a fear that some of those roads would direct people into the fire. Evacuees were herded onto Front Street, where traffic was at a standstill. Some people abandoned their vehicles and hurled themselves into the ocean. The water's surface itself seemed to be smoking, making it hard to breathe. One group held on to wreckage that had fallen in the water; others waded for hours, trying to dodge or douse the embers falling on their heads.

By 7 p.m., the docks and boats in the harbor were lit up as if in a coalfired oven, the roar of the flames broken by a staccato of exploding propane tanks. In the ocean, the current was pulling weaker swimmers out to sea. Coast Guard boats were crisscrossing the water, barely able to see through the smoke. They ultimately rescued seventeen people from the water and forty from the shore, and recovered one body the next day.

During the fire, the county's command-and-communication system fell apart. The county sent one emergency cell-phone evacuation alert at 4:16 p.m., after the fire was already moving through town, but the order was just for a single neighborhood. At

6:03 p.m., while the fire was incinerating Front Street, and while people were struggling in the sea, Maui County's mayor, Richard Bissen, appeared on a local news broadcast, calmly sitting in his office on the other side of the island. "I'm happy to report that the road is open to and from Lahaina," he said, seemingly unaware of the inferno under way. The county did not issue online evacuation orders for other parts of town until 9:45 p.m. The winds finally subsided at dawn.

Maui was formed by two shield volcanoes about two million years ago, becoming the second-largest island in the Hawaiian archipelago, the most remote chain of inhabited islands on earth. Lahaina, which means "cruel sun," sits on the leeward side of Maui, below the western mountains, Mauna Kahālāwai, which roughly translates to "house of water." The highest peak is one of the wettest places in the world, historically receiving about three hundred and sixty-six inches of rain per year.

Hawaiians built their communities around the watershed. Their word for water, *wai*, has many meanings: blood, passion, life. Lahaina—even though it was relatively hot and dry—became, because of its water supply, a cornucopia, replete with irrigated breadfruit, banana, and sugarcane crops, terraced taro patches, and fishponds. In the early nineteenth century, Lahaina was the capital of the Hawaiian Kingdom. The king lived in a coral-block palace on an island in the middle of a pond. Residents could paddle around town.

During the American Civil War, the agricultural economy that sustained Southern farmers collapsed, and Hawaii became a primary source of sugar. But sugarcane is a thirsty crop. One ton of sugar requires a million gallons of water. To meet that demand, private companies producing sugar (and, later, pineapples) rerouted the flow from Maui's watersheds, building concrete ditches, tunnels, pipes, flumes, siphons, and trestles across the island. European ranchers introduced non-native, drought-resistant African grasses—guinea, molasses, and buffel—for grazing livestock. In less than five decades, the island's landscape and ecology were dramatically altered.

Agriculture declined in the late twentieth century, and plantation owners abandoned vast swaths of farmland, allowing the non-native grasses to proliferate. Instead of restoring the steep mountain streams, they left their diversions in place—in some cases, dumping water into dry gulches, or directly into the ocean—or used them to develop beachfront resorts, with lush gardens, swimming pools, and golf courses. By 1996, as Carol Wilcox writes in her chronicle *Sugar Water*, "competition for water had met the limits of the resource in Lahaina." That same year, the newly formed West Maui Land Company started buying abandoned plantations (and their valuable irrigation systems) and creating new subdivisions.

Natural wildfire on Maui used to be rare. The high-elevation endemic forest acted like a sponge—capturing fog and rain, recharging aquifers, and releasing water downstream. But land development and the encroachment of invasive species are shrinking this ecosystem. "Towns are now, instead, surrounded by tinder-dry invasive grasses that just go up in an instant," Carter told me.

In the past decade, Maui has faced periods of severe drought, exacerbated by climate change. Parts of the island got so dry during the past two years that the county limited residential water use. Hotels did not face restrictions. Fodor's Travel included Maui on its 2023 "No List," which warns against visiting regions that are suffering from environmental threats. And yet tourism in Maui remained steady.

Native-Hawaiian-sovereignty groups have long been fighting for stream restoration and more water control. According to the state constitution and a series of landmark court cases, Hawaii's water must be held in a public trust for the people's benefit, which includes the use of water for traditional and customary practices, such as taro farming. Private developers are required to follow streamflow standards, and must get approvals from the state's Commission on Water Resource Management if they want to divert more water than their usual allotment.

On August 10, as fires in Olinda continued to burn, the governor, Josh Green, suspended the water code. The same day, Glenn Tremble, a partner at the West Maui Land Company, wrote a letter to the water commission stating that on August 8 he had asked to divert stream water to the company's reservoirs, south of Lahaina, to help put out the flames. A water commission deputy director named M. Kaleo Manuel delayed the diversion until that evening, explaining that Tremble first needed to check with a downstream taro farmer who relied on the stream to fight fire on his property. That stream is not connected to the county's water network, which supplies Lahaina's fire-hydrant system. Moreover, the day's heavy winds meant that helicopters could not use those reservoirs to fill water bombs (known as Bambi Buckets)—they could not fly at all. Still, many were eager to blame the Native Hawaiian water deputy and, by extension, the water code. (A headline in the *New York Post* read, "Hawaii official concerned with 'equity' delayed releasing water for more than 5 hours as wildfires raged.") Manuel was reassigned to another department.

Peter Martin, West Maui Land's cofounder and CEO, told me that protecting water for Native Hawaiian cultural practices was "a crock of shit," and that invasive grasses and "this stupid climate-change thing" had "nothing to do with the fire." He felt unfairly demonized by activists: "They're trying to paint this picture that I'm a colonialist." The real problem, he said, was the water commission and its code, which was so over-bearing that it prevented him from replacing dry grassland with irrigated, landscaped parcels, or even small hobby farms. Maui's lands, he added, "weren't being used as God intended."

Sixteen days after the fire, Maui County, with help from the FBI, released a list of three hundred and eighty-eight missing people. This was a distillation of a larger list, with more than a thousand names, that had been assembled from potentially unreliable sources—online groups, anonymous calls—and contained redundancies and errors. (Many individuals had the strange experience of seeing the list and learning that the FBI. thought they were missing.) The estimated death toll had remained the same since August 21, when the police announced that they had recovered a hundred and fifteen bodies.

Two blocks from Saribay's house, Alfredo Galinato, a seventy-nine-year-old Filipino immigrant, had lived with his wife, Virginia, and their son James, who is mentally disabled. When the fire approached, Galinato told James to run to Safeway, where Virginia worked. Then Galinato climbed onto his roof with a hose to soak the house, just as he had done during previous fires. Virginia and James survived, but Galinato was now among the missing. His two other sons, Joshua and John, who were not in Lahaina on August 8, went to the burn zone to look for their father. "Everything was burned to dust," Joshua told me. After searching for seventy-two hours, they heard that authorities were collecting DNA samples from people with missing relatives, and they went to a community center to get their cheeks swabbed.

Following the fire, forensic anthropologists, dentists, pathologists, and fingerprint and X-ray technicians flew to Maui, to aid the overwhelmed coroner's office. Urban search-and-rescue teams, deployed by FEMA, started working with cadaver dogs across the five and a half square miles that had burned.

The Lahaina fire reached temperatures more than a thousand degrees hotter than the temperature on Venus. Four thousand vehicles were caught in the flames, and almost none of them were left with tire rims. "There were rivers of melted aluminum down the streets," Stephen Bjune, the spokesperson for FEMA's Urban Search and Rescue Team, told me. But the fire also moved in mysterious ways. A truck on Front Street had been full of glass bottles for recycling, all of which melted. Five feet away, a single silver minivan was unmarred, as if it were still sitting in traffic.

It was so hot in the burn zone that the dogs could work only in quarter-hour shifts. About 15 percent of the discovered remains were intact enough to obtain fingerprints from—that is generally the quickest route to identification. In another 13 percent, forensic dentists were able to identify people from their teeth. In 2 percent, medical hardware—such as a pacemaker—was used to make identifications. But, for about 70 percent of the victims, the experts needed DNA. In the majority of those cases, there were still significant amounts of tissue. In a few cases—the most difficult ones—there were only ashes and small fragments of bone.

The DNA analysis was conducted with the help of ANDE Rapid DNA, a biotech and public-safety company. ANDE manufactures a hundred-pound printer-size instrument that can generate a DNA profile, or "fingerprint," in two hours. It can analyze five samples at a time—drops of blood, pinhead-size bits of liver, or fragments of bone. Richard Selden, the company's founder and chief scientific officer, said that he and his team initially developed the instrument for the United States military's counterterrorism operations in the Middle East, so it was designed to be portable and rugged.

The DNA fingerprints were compared with reference samples that families, like the Galinato brothers, had provided. The problem was that many family members were not submitting samples. Some authorities attributed this to a lack of trust between residents and the government, which went back more than a century, to colonization. Officials launched a publicity campaign emphasizing that the DNA samples would be used only by ANDE, and would not be used by the government for tracking people.

Alfredo Galinato was one of the first victims to be identified using the rapid-DNA machine. He had worked as a groundsman at the Westin, near Lahaina, for twenty-five years, and loved taking care of the hotel's parrots. I met his family across the island, at the house of his son John's fiancée, about a week after they received confirmation of his death. John, a carpenter, looked just like his father—with a gentle, open face and the strong, scrappy build of a former high-school state wrestling champion. He said that he felt blessed to have found out about his dad relatively quickly, compared with all those who were still searching.

Later, as I was driving back to Lahaina, John sent me a text, written as if his father were still alive. "Idk if I mentioned. My dad is a hard-working man, dependable," he wrote. "I can count on him."

A strange scale of tragedy had developed on Maui. Those who hadn't lost loved ones might still have lost everything they owned. And yet some said they felt lucky. "Just material things," one person told me. A woman named Michele Pigott, who had lived in Lahaina since 2011, said that this was the third house she had lost to a fire. (The first two were in California.) She was almost immune to being displaced again. "Piece of cake," she told me. "There's not a goddam thing you can do."

But anger was pulsing under the surface. As one mother said to me of the disaster response, "How could so many people fail at their job at the same time?" Among the first failures were the warning sirens. Although Maui has eighty of them, none were activated when the fire began. A week after the disaster, Herman Andaya, the administrator of the Maui Emergency Management Agency, defended his decision not to use the sirens, saying that they were primarily for tsunamis—even though the agency's Web site lists brush fires as one of the reasons for the "all-hazard siren system" to go off. Andaya said he had been concerned that the sirens would send people fleeing to higher ground, into the flames. He also said he was afraid that people wouldn't even hear the sirens, because almost all of them are along the coastline, and that he did not regret his decision. The following day he resigned, citing health reasons.

Another problem was the lack of firefighters. The Maui Fire Department has long been short-staffed and underfunded. Despite the vast increase in wildfire country on the island, the last time a new station was built was in 2003. West Maui's population has grown from roughly eighteen thousand to twenty-eight thousand over that span, and is serviced by two stations and three trucks. No more than sixteen firefighters were initially on duty in Lahaina on the afternoon of August 8. "They did an extraordinary job," Bobby Lee, the president of the Hawaii Fire Fighters Association, told me—"before they ran out of water." County water levels were already low, and then the fire hydrants lost too much pressure. Some ran dry. The fire's extreme heat had caused water lines to break, something that also happened in a catastrophic urban fire in Fort McMurray, Canada, in 2016.

Many survivors have said that they received no evacuation orders from the police. When Mayor Bissen was later asked why, he said that, in fact, police officers had driven the streets, calling from loudspeakers. But that had happened later in the evening, near where fires were still burning on Lahaina's north end. After a local reporter pressed him on the failure, Bissen said, "You can decide what the reason was, whether it was somebody did something on purpose, or somebody did something out of negligence, or somebody did something out of necessity. There are probably a lot of reasons you can apply to why we do what we do as human beings."

I asked the Maui County police chief, John Pelletier, about all the roads out of Lahaina that had been closed. He said, "There was always a way out, if people were willing to go that way. Nobody was barred from going out of Lahaina town." He continued, "We were encouraging everybody to get out, but it just depends on the dynamic. It may not have been the way that they maybe wanted to go."

The nature of the disaster, and the chaos and information void in the aftermath, lent itself to rumor and conspiracy theories. Selden, the ANDE scientist, told me that there are two kinds of disasters: open and closed. A plane crash is the latter—there is one site of wreckage and a manifest listing who was on board. Lahaina is open. There is no list of people who were in town that day, and the burn area is large and unfixed. Speculation about the demographics of the victims was rampant. Because school was not starting until August 9, people thought a lot of children might have been home. As of late August, only two families had reported the loss of a child; the police had not confirmed their deaths or identities.

Keyiro Fuentes, who was fourteen years old, was at home, asleep with the family dog, when the fire swept onto his street. His mother came back from work to get him, but the police blocked her, saying that they had already cleared the area. Days later, the family found Fuentes's body in the house. His father wrapped the body in a tarp and, with his older son's help, drove Fuentes to a police station. "The first thing I said was, 'Mr. Officer, I have a body and it's that of my little brother,'" Josue Garcia Vargas, Fuentes's twenty-year-old brother, recalled. One officer at the station seemed to be in shock. "His hands were shaking," Vargas said. "I kept telling him the name, and he kept saying, 'What? What?'"

In mid-September, the police confirmed Fuentes's death. The identification was delayed because Fuentes was adopted, and the police had to obtain DNA samples from his biological family, in Mexico, to confirm that he was who the Vargas family said he was. But the Vargases had already held a memorial. A week after the fire, when Fuentes would have turned fifteen, his mother threw him a birthday party.

Months earlier, Fuentes had told Vargas about a girl he had a crush on in his class. She hadn't seemed interested, so Vargas suggested that Fuentes flirt with the girl's cousin to make her jealous. Both girls had attended the memorial. "They were both crying, man," Vargas told me when we met, tears rolling down his face, although he was smiling. "He made them cry. That made me happy." Fuentes had been a tough, fiery, and sweet little kid, who loved mixed martial arts. "He wanted to be a police officer," Vargas reminisced. "He saw when my mom got screamed at by one of our neighbors and he got mad and said, 'I'm going to be an officer so this will never happen to you.'"

When the family had found Fuentes, Vargas added, their dog's remains were there, too. "We think they were hugging each other," he said, now hugging himself, struggling to speak. He reminisced, of his brother, "He was always there, making his presence known, saying 'Wassup, bro!'" He paused. "It's hard for me to accept the reality of what happened."

On August 29, Pelletier announced that recovery crews had completed 99 percent of their land search in Lahaina. More than three hundred people remained unaccounted for, but the estimated number of deaths had not changed. The Galinatos' neighbor, a forty-three-year-old EMT named Tony Simpson, was still missing. The day of the fire, Simpson's parents were at home in Belize, his sister Nichol was in Thailand, his other sister, Nova, was in Connecticut, and his brother was in New York. After a couple of days, none of them had heard from Simpson, and they started to panic. They made dozens of calls—to his employer, to the Red Cross, to the FBI, to the police. Nichol posted Simpson's photo in a Maui-disaster-relief Facebook group. Nova filed a missing persons report and submitted a DNA sample to an FBI office in Connecticut.

The family had agreed that it made sense to do what they could from a distance, rather than get in the way of the authorities. But, after two weeks, Nichol and her husband, Angel Priest, made the forty-hour journey from Thailand to Maui. Their first stop was the Family Assistance Center, which was housed in a Hyatt Regency hotel. The complex was full of displaced people wandering a maze of courtyards, shuttered shops, and gardens. Nichol sat to give a DNA sample; ANDE had an instrument on-site. She asked if her sister Nova's DNA was already in their system. The workers didn't know.

Nichol soon learned that a large percentage of victims were recovered within a few blocks of Simpson's home. When she told other families on the island where her brother lived, they'd offer condolences. Nichol tried to visit Simpson's house in the burn zone but was stopped by the National Guard and told she needed an official escort. She called the police, and a receptionist suggested that she call the EOC. When Nichol asked what the EOC was, the receptionist didn't know. (EOC is the Emergency Operations Center.) Nichol reached a person at the EOC, but learned that she could not, in fact, get an official escort into the burn zone. She was also told, by a field worker, that the residential area had been fully searched. That is, except for multistory buildings. This only confused her more. Simpson's house was two stories. Had it been searched? Unclear.

Nichol and Priest talked to unsheltered people in encampments. Simpson had a strong tie to that community; he had moved to Maui with a friend who chose to live outside. "We're literally stopping people on the street and asking them, 'Do you live here? Can you help us find a place to go search?'" Nichol told me. "Tomorrow we're going to find some random cave somebody suggested." I asked if they really believed that Simpson was hiding out some-where. "Absolutely," Nichol said. Simpson had led an eclectic life. He lived off the grid for two years, "on mangoes, basically, like a friggin' fruitarian." She added, "We could just see him showing up later with some crazy story." Like he'd been living in a cave for two weeks. "Maybe he can make a big Hollywood movie about it," she said, letting out a belly laugh. "Actually, he would hate that."

After ten days, Nichol and Priest decided to fly to Belize to be with her parents. Before they left, they drove to the Lahaina post office to get Simpson's mail forwarded to them. "We were really grasping at straws for small things that I could take back to my family," she told me. "Because we have nothing of his." This was true of many victims' families. The Galinatos had lost most of Al-fredo's belongings, although his wedding ring had been recovered by search-and-rescue workers.

On the way to the post office, Nichol received a call from the Maui P.D. The police had matched her DNA with her brother's remains, which they had found on August 11—twenty-one days earlier—in a burned structure near his house, along with the re-mains of several others. As people had sought shelter, they landed in others' homes, businesses, or cars, and in some cases died to-gether. The location and commingling of remains delayed the processing of samples, and comparisons with the families' DNA. This situation also resulted in an initial overcount of victims; two different body bags might later have been found to contain one person.

Nichol was not only heartbroken by her brother's death but frustrated by the lack of clear communication from the author-ities. "We're thinking, They've recovered a hundred and fifteen bodies. They've recovered no more in several weeks. We don't match any of those bodies, so Tony must still be missing," she told me. "It brought us a lot of false hope."

The morning after the fire, when Saribay was leaving the church parking lot, he saw smoke in the direction of a house belonging to his kids' grandparents, in a neighborhood called Leiali'i. He drove there and found his brother, who told him that another house, bordering their friend Archie Kalepa's property, was smol-dering. The fire department had already been there, but the fire had flared back up.

Saribay and his brother ran across neighbors' gardens, grab-
bing more hoses. They broke Kalepa's fence and soaked his yard.
Saribay's shirt had melted the night before, but he'd found a
backpack containing women's clothes that he had changed into.
Saribay has a mischievous streak, which, despite what he had
been through, hadn't gone away. "I fucking fought that mother-
fucker while I was in a red fuckin' blouse," he said.

They extinguished the fire. Leiali'i had been built seventeen
years ago, as part of the Hawaiian Homes Commission Act, which
allots homesteads to people who have at least 50 percent Hawaiian blood. And the neighborhood was saved. Of its hundred and
four houses, only two burned down. "At 10:30 p.m., when I evacu-
ated, the flames were as high as the trees right behind my house.
I thought it'd be gone," Rodney Pa'ahana, the president of the
Leiali'i Association, a community group, told me. "We were as-
tounded," he continued. "God put a finger on us, as if to say, The
Hawaiian people need to stay and rebuild."

Kalepa had been in California during the fire, but he came home
on the first flight he could. When he arrived, Saribay apologized
for breaking his fence. "Fuck my fence!" Kalepa told him. "You're
the guy who saved my house!" Within forty-eight hours, that house
became one of Maui's first community-organized emergency hubs.
Kalepa told me that the donations, which ranged from money to
food and supplies, had been overwhelming. People were sending
poi—a traditional Hawaiian staple consisting of paste made from
ground taro—from four islands away. Lahaina residents started calling the house "the local Costco."

I met Kalepa on the cul-de-sac outside his house in late August,
under a cluster of pop-up tents. There were more than two dozen
coolers, towers of water bottles, Clorox wipes, a bleeping Starlink
router (for Internet), and a machine that converted moisture
from the air into water. Friends and volunteers were lugging boxes
and ice, setting up rooftop solar panels, and peeling bananas to
make banana bread.

Kalepa asked if I wanted to see the line where Saribay had held
off the fire, gesturing toward the back yard. It all seemed fairly
normal. But at the edge—beyond the grass, palm fans, magenta
stalks, and yellow frangipani flowers with pink centers—there was
a gap where the fence had been. On the other side of that gap,
the world was suddenly black-and-white. A foundation of scorched

cinder blocks suggested the ghost of a house. Rusted rebar poked the air. There was a shovel, bent like a bow tie. A hollowed pickup truck was snapped in half. The air was stagnant with the lingering, acrid smell of smoke, rot, and death.

Kalepa, a ninth-generation Hawaiian, recently turned sixty. He is a former lifeguard and big-wave surfer who provides ocean training to Navy SEALs. In the weeks since the fire, he has become one of Maui's most prominent community leaders. "I never wanted to be in this position," he told me. "I was really enjoying my life." He is now serving on Mayor Bissen's five-member Lahaina Advisory Team, which will consult on the town's rebuilding. "We have one chance at fixing this," Kalepa said. "And, if we get it wrong, all of Hawaii's going to fail. Not just Lahaina." One of the community's biggest fears is that the process will favor developers, tourists, and the wealthy. Kalepa, other activists, and water-rights groups have been strenuously advocating for the local community. On September 8, Governor Green announced that he was reinstating the state's water code. Several weeks later, the water deputy, M. Kaleo Manuel, was returned to his post.

An organization called the Fire Safety Research Institute has been selected to investigate the government's response to the catastrophe. Initial findings are expected by December. But responsibility for the fire falls in many places, on many individuals, across the decades. "For the last hundred and fifty years," Kalepa said, "Hawaii's gone in the wrong direction. This situation we're in right now? It brought that to light."

People often view disaster survivors' stories as they would an apocalypse film—a frightening but faraway and anomalous event, witnessed from a safe place. But these stories are missives from our immediate future—postcards from what, one day, might be your circumstance, in this era that some climate-change experts now call the Pyrocene. Record-breaking wildfires are happening more frequently all over the world, with studies directly linking climate change to the increase in fire duration, size, and severity. Wildfires in the U.S. caused more than eighty billion dollars in damage from 2017 to 2021, a nearly tenfold increase from the previous five years.

Hawaii has made gestures at addressing climate change; in 2015, it was the first state to pledge to convert entirely to renewable energy by 2045. And yet critics have jumped to blame the power company, Hawaiian Electric, for focusing on renewables, claiming that it was doing so at the expense of maintenance that could have prevented the West Maui fire. Similar debates are playing out all over the country, where the same funds required for infrastructure maintenance and improvements, in this hot new world, are also needed for the green-energy transition.

Hawaiian Electric's CEO, Shelee Kimura, testified at a congressional hearing that thousands of aging utility poles had not been tested for termites or rot since 2013, but she also said that power lines had been de-energized for more than six hours before the afternoon fire began, and that the company was therefore not responsible. Her assertion, and the fire's true cause of ignition, are under investigation, and the company now faces more than a dozen lawsuits, including one filed by Maui County.

On October 8, the two-month anniversary of the fire, Governor Green welcomed tourists back to parts of West Maui. Many community members were outraged; they felt that they weren't near ready. Just a few days later, more human remains were found in Lahaina. Six people are still missing, and there is one body that has not yet been identified. "Imagine what happens when you gotta live in temporary housing, surrounded by ash, and go to work back in those hotels," Nāʻālehu Anthony, a filmmaker and an activist, told me. "People just hit this wall where they're saying, 'We're not going to do that anymore.'"

Even though returning to work was hard to stomach, it was crucial to Maui's economy, which is heavily reliant on tourism, and necessary for residents, who were struggling with bills and insurance. Many residents were worried about their mortgage payments, which are still due even after your house burns down. "For what, a piece of dirt?" Saribay said. His kids were O.K., which was "all that matters," he said, but he had lost three of his houses, his tattoo parlor, and his boat. He was living in his kids' grandparents' house in Leialiʻi. Saribay told me that he had taken a forty-hour course to obtain a hazmat certification, so that he could be part of the effort to clear the rubble from Lahaina. But the idea had become a nightmare. "I just don't want to be *in there* right now," he said.

"Fifty to sixty percent of the people that passed away was from my neighborhood," Saribay told me. He has been dealing with trauma: "My nights are a fucking question mark," he said. "I'm so tired. My mind races." He has thought about leaving Hawaii altogether, and has felt financial pressure to sell his land—a common experience among homeowners, some of whom reported receiving calls from real-estate investors just days after the fire. "I could just be outta here and say, 'Fuck Hawaii,'" Saribay continued. "I'm not gonna, but fuck."

He has had delays with his FEMA relief application—he still doesn't know how much money or what kind of housing assistance he will get. "Everything will be O.K. if the government really helps us, but they're not," he said. "It's the people of Maui who's helping each other."

One Friday evening, I attended a community meeting at Kalepa's house, which had become a weekly event. People offered advice, consolation, ideas. One man discussed new air purifiers that had been donated by a nonprofit, which residents could take home with them. Pa'ahana, the Leiali'i Association president, gave a teary speech arguing that Lahaina should be rebuilt as a giant beach park, with all the shops and homes staying up near the highway. "I know I'm gonna get a lot of flak from the billionaires and businesses," he said. "But, if we do this right, they will thank us when we're not here anymore." As he spoke, fat raindrops started falling. Kalepa told the crowd, "The blessings are pouring out for us."

Many Hawaiians want to make this moment an opportunity. "It's very rare to have people plan a new town after hundreds of years of history," Pa'ahana told me. "But we get a chance." The tropical shower stopped as suddenly as it had started. A line of volunteers carried platters of opakapaka, venison, coconut, and poi to folding tables set up in the cul-de-sac. Saribay was bopping around, taking pictures of the food and cracking jokes. "He's so full of life," Kalepa said, grinning in his friend's direction. As it got dark, kids sat on the asphalt playing duck-duck-goose. Anthony, the filmmaker, told me, "The reason Archie Kalepa stood this up is because his community needed help, and because the idea of aloha is not how much you can keep. It's how much you can give away."

The first week and a half after the fire, apart from the machinery and the dogs, Lahaina was silent. No birds or bugs were alive. But even among the ashes there is virescence. The oldest banyan tree in Lahaina, planted a century and a half ago, beaten and blackened by fire, has sprouted green buds. They appear to glow against the surrounding moonscape, like time travelers from our once and future planet.

IAN FRAZIER

Deep in the Wilderness, the World's Largest Beaver Dam Endures

FROM *Yale Environment 360*

WOOD BUFFALO NATIONAL Park, the largest national park in Canada, covers an area the size of Switzerland and stretches from Northern Alberta into the Northwest Territories. Only one road enters it from Alberta, and one from the NWT. If not for people observing it from airplanes and helicopters, and satellites photographing it, little would be known about big parts of it. The park is a variety of landscapes—boreal swamps, fens, bogs, black spruce forests, salt flats, gypsum karst, permafrost islands, and prairies that extend the continent's central plains to their northern limit. The wood buffalo in the park's name are bison related to the Great Plains bison. In this remoteness, the buffalo descend from the original population, and the wolves that prey on them are also the wild originals. Millions of birds summer and breed here. The park holds one of the last remaining breeding grounds of the whooping crane.

Other superlatives and near-superlatives: the delta in the park's southeast where the Peace River and the Athabasca River come together is one of the largest freshwater deltas in the world; last summer, some of Canada's largest forest fires burned in the park and around it; and—just inside the park's southern border—is the largest beaver dam in the world.

The dam is about a half-mile long and in the shape of an arc made of connected arcs, like a recurve bow. The media has known about it for sixteen years, and in that time no bigger beaver dam has come to light, so it's still known as the biggest, and scientists believe it almost certainly is. Animal technology created it, but human technology revealed it. In 2007, Jean Thie, a Dutch-born landscape ecologist who lives near Ottawa, was looking at the latest satellite imagery of places he had examined via satellite in 1973 and 1974, when he was studying permafrost. It's hard to remember, but in the early '70s some scientists thought the Earth might be cooling. Thie's research had showed evidence of the opposite; the paper about permafrost melting that he published in 1974 is now considered one of the pioneering studies of climate change.

As he looked over 1970s images taken by NASA's Landsat satellite and compared them with the latest images from Google Earth and other sources, he noticed that in certain landscapes the evidence of beavers now was everywhere. From being almost wiped out by the fur trade between about 1600 and the twentieth century, beavers had bounced back. Just one example was a belt about 1,100 miles long that extended into Wood Buffalo Park. Among the hundreds of beaver dams in this area, Thie came across one that looked bigger. He measured it and found it to be 2,790 feet long, or about a half-mile. The 17-acre lake created by the dam reposed undisturbed, shiny and opaque in its swampy northern forest, and in the middle of the lake the small brown dot of a beaver lodge could be seen.

On October 5, 2007, Thie posted the satellite photo of the dam on the Google Earth Community Forum, with text explaining that it was probably the world's largest. Seven months later, a reporter for Canadian Broadcasting Company Radio saw the posting and did a story about it. Other outlets picked up the story and "the world's largest beaver dam," a phrase that's satisfying to say and think about, achieved a modest international fame.

Many of the beavers that have reestablished themselves globally are descended from beavers that were planted by wildlife biologists. The thriving beaver population of Tierra del Fuego (another place Thie has studied) is descended from beavers brought to Argentina from Canada's Saskatchewan River, who are themselves scions of beavers transplanted from upstate New York. No reintroduction of beavers was done in Wood Buffalo Park. Thie believes that the beavers who built the dam are of original stock. Like the wood buffalo and the wolves, they were too remote to be wiped out.

The officials who run the park heard about the world's largest beaver dam because of Thie's discovery, like everybody else. Until the CBC reporter called them for comment, they had not known that their park contained the world's largest beaver dam. None of the park's personnel had ever been to it, or has visited it on the ground (or what passes for ground there), to this day. When I called Tim Gauthier, the park's external relations manager, he said that he had flown over the dam many times but never stood on or near it. He did not know if the water in the lake was still deep enough to cover the entrance to the lodge or lodges. In these remoter areas of the park, he said, "we tend to let such things regulate themselves."

Since 1983, Wood Buffalo National Park has been listed as a World Heritage Site by UNESCO, the environmental and cultural agency of the United Nations. In more recent years, this designation has become shaky; at the request of the Mikisew Cree First Nation, whose members gather traditional resources in the park and depend on it for cultural survival, UNESCO has twice investigated environmental threats to the park and has come close to declaring it officially endangered. Wood Buffalo Park is now on UNESCO probation, and the governments of Canada and Alberta are supposed to fix its problems.

The park is suffering the worst drought in its history. Flows are down by half in many places, owing to climate change, water diversion, poor seasonal snowpack, and dams on the Peace River, upstream in British Columbia. A danger that seems inescapable comes from the oil sands that are being mined for crude-oil-containing bitumen, and from tailing ponds that hold trillions of liters of mine-contaminated water. The ponds are near the banks of the Athabasca River, just upstream from the park boundary. They are fatal to birds that land on them. Given the direction that water flows, conservationists and native people fear the tailings will pollute the park eventually. Toxic chemicals have already been found in McClelland Lake, just southeast of the park. Locals stopped taking their drinking water from the lake years ago.

Gillian Chow-Fraser, the boreal program manager for the Northern Alberta chapter of the Canadian Parks and Wilderness Society, in Edmonton, travels in the park often by helicopter, canoe, and foot. She has described the park's environment as "super degraded." When I spoke with her by phone not long ago, she talked about a recent tailing basin leak that was not reported to the First Nations downstream of it for nine months. In places that used to flood regularly but now don't, the land is drying out and vegetation disappearing. Though she crisscrosses the park, she has never seen the world's largest beaver dam, but she's grateful that it's there and bringing the park attention.

Another expert, Phillip Meintzer, conservation specialist with the Alberta Wilderness Association, told me that he hadn't seen the dam, either, but that the park's difficulty of access is a good thing, in a way, because it keeps people from visiting in large numbers and putting stress on the area. The downside is that environmental degradation, like the recent tailings seepage, can happen without many watchers finding out. Meintzer's main worry is that when the economy shifts to renewables, the oil sands will be abandoned and taxpayers stuck with the cleanup. What will be done about the multi-trillion liters of toxic tailings is unknown. "Last summer I was on a trip to test water quality in and around McClelland Lake," he said. "We camped by the shore, and all night in this remote and uninhabited place we could hear the propane cannons at the nearby tailings ponds firing to scare off the birds."

As far as is known, only one person has ever been to the world's largest beaver dam. In July 2014, Rob Mark, of Maplewood, New Jersey, forty-four years old at the time, reached the dam after a challenging journey. Holding the flag of the Explorers Club, the international organization with headquarters in New York City, he took a photo of himself standing on the dam. The top of the structure was the only solid ground he had encountered for miles. After he got back, a newspaper in Edmonton did a story about him, and he appeared in other newspapers and a travel magazine. His achievement is like the dam in that so far no one has said it isn't unique.

Mark is now a blueberry farmer in Virginia. When I reached him by phone, he told me he did solo extreme treks unsponsored and for his own pleasure. In 2007, he crossed South America from the Pacific Ocean to the Amazon River by hiking over the Andes. The idea of going to the world's largest beaver dam occurred to him after he read about Jean Thie's discovery. He planned the trek for several years, and in 2011, he flew to Fort McMurray—the Alberta town, more than 100 miles from the dam, that is the hub of the oil sands industry—to see how he could get from A to B.

His plan was to go down the Athabasca River by boat, then hike through the muskeg peatland. That proving impracticable, he returned home and decided he would come at the dam from another direction, by way of Lake Claire, whose southwestern edge is about 10 miles from it. Crossing the lake by boat, a distance of about 25 miles, and then hiking to the dam, seemed straightforward enough. But the lake is more like a wetter spot in a swamp than a lake. Sometimes it does not have enough water for boats. Mark waited three years for that problem to improve. In 2014 it did, and Mark went to the town of Fort Chipewyan, east of Lake Claire, and hired a man to ferry him.

The lake has no real shore, it just gets shallower at the edges. At a chosen point Mark got out and arranged for the boatman to return and pick him up there in six days. Mark noted the coordinates in his handheld GPS and told them to the boatman. The boatman replied that he had no GPS. That was a detail Mark had not thought of. The boatman told him to cut one of the nearby willows and stick it in a more conspicuous place in the swamp-lake, and they arranged to meet by it. Then the boatman left, and Mark began his trek.

The mosquitos swarmed like nothing he'd seen in the Amazon. He was ready for that and for trying not to go crazy from their noise. The sphagnum moss islands submerged slowly under his weight, step by step, as he grasped at willows to sort of brachiate on. By looking at the tree species shown on satellite photos he had plotted a route along comparatively higher ground, and he tried to keep to that. Mostly the route, which required two days of slogging, was just swamp. The last mile to the beaver dam took him five hours.

Late in the long subarctic afternoon he emerged into the clear patch of sky created by the dam's lake, waded to the dam, and stepped onto it. The dam is no more than three feet high at any point. He realized that a person seeing it up close would never guess it extended for half a mile. To grasp its full size and the ingenuity of its construction you needed a photograph from space. A lone beaver appeared, looked at him, and slapped its tail. Mark got a sense that his presence enraged the beaver.

Bringing out his Explorers Club flag from his pack, he took the selfie. To be allowed to carry that flag he had had to apply to the club, which reviewed his plan of exploration and deemed it worthwhile. Mark became the 851st explorer in the club's 110 years to carry the flag, joining a list that includes Thor Heyerdahl and James Cameron. After a supper of granola and peanut butter, he hiked to some larger spruce nearby, lashed his hammock between two of them, draped the mosquito netting, and prepared to spend the night.

Hiking out occupied three more days. When he reached the lake, he could not wait next to the willow marker for his ride, because that would mean standing thigh-deep in water. He sat on a drier patch of ground back in the trees, too far from the lake to see it, and listened for the engine. At mid-morning of the day appointed, he heard a sound that got louder. The boatman went right to the unlikely willow and Mark walked through swamp to the lake and waded out to the boat, so exhausted he could barely climb in.

The world's largest beaver dam is not like human dams. It does not stopper a river, or even a stream or rivulet. Its low half-mile barrier collects small trickles that come off a plateau called the Birch Mountains. Along the margin of this comparatively higher ground, it accommodates itself to a slope of less than 2 percent. The gathered-up trickles have amounted to a lake, and after the beavers eat the plants that grow in it, they may relocate to another dam and another pond, graze that area, then move on again, in a sort of crop rotation. Other dams in this beaver belt are up to three-quarters the length of the longest dam. These long, low dams may help the beavers adapt to drought.

Places almost impossible to get to undergird all of existence. In my car there are regions under the front seats where, when my cell phone falls into them, I must almost take the car apart to get it out. Beavers create hard-to-access places that are good for them, less so for us. Jean Thie had beavers on land he owned near Ottawa, and they built dams and made swampy ponds and cut down trees. He got a trapper to remove them but they or other beavers came back. Finally, he gave up and just put chicken wire around the trunks of trees on the property and lived with the beaver landscape.

In the big picture, Thie is pro-beaver nonetheless. "Of course, I'm not very positively minded about our own future on the planet," he told me. "But I am an optimist about beavers. Their presence improves water management, reduces water flows, reduces the loss of runoff, and creates and improves wetlands. In drier landscapes of the future all this could be of benefit. I think the worldwide flourishing of beavers is a small step in a good direction."

BEN GOLDFARB

City of Glass

FROM *bioGraphic*

EVERY SPRING, AS the daylight lengthens and the weather warms, rivers of birds flow north across the Midwest. They fly high and at night, navigating via the stars and their own internal compasses: kinglets and creepers, woodpeckers and warblers, sparrows and shrikes.

They come from as far as Central America, bound for Minnesotan wetlands, Canadian boreal forests, and Arctic tundra. They migrate over towns and prairies and cornfields; they soar over the black tongue of Lake Michigan in such dense aggregations that they register on radar. Upon crossing the water, many encounter Chicago, where they alight in whatever greenery they can find—office parks and rooftop shrubs and scraggly street-trees and the sparse landscaping outside apartment-complex lobbies.

And, as they linger and forage in Chicago's urban canyons, they collide with glass.

To us humans, glass is ubiquitous and banal; to birds, it's one of the world's most confounding materials. A tanager or flicker flying toward a transparent window perceives only the space and objects beyond, not the invisible forcefield in its way. The reflective glass that coats many modern skyscrapers is just as dangerous, a shimmering mirror of clouds and trees. Some birds survive collisions, dazed but unharmed. Most don't, done in by brain injuries and internal bleeding. Per one 2014 analysis, glass kills as many as a billion birds every year in the United States alone.

Chicago, among the largest and brightest cities within North America's Mississippi flyway, is especially lethal—both during spring migration and again in fall, when the survivors fly south. The millions of artificial lights that glow across the Windy City present as a galaxy of false stars, confusing migrant birds that orient themselves by starlight and enticing them toward the glassy buildings below. In 2019, researchers at the Cornell Lab of Ornithology ranked Chicago the country's most perilous city for birds—a metropolis that doubles as an ecological trap.

The city's residents aren't blind to the tragedy. Some architects and building managers have taken measures to protect birds, and politicians have tried to alleviate the crisis through laws and regulations. But progress has been fitful, and new glass monoliths sprout every year. Chicago thus epitomizes both the severity of the United States' glass problem and the difficulty of summoning the will to redress it. "We have so much urban lighting, so much glass, it just puts all the wrong things together for birds," Annette Prince, the director of a conservation group called the Chicago Bird Collision Monitors, told me. "Chicago is the perfect storm."

One morning at the outset of spring migration, I found myself pacing Federal Plaza in downtown Chicago, waiting to join Prince as she scoured the city for birds. I shivered in the predawn damp, the sky the pearly gray of a chickadee's wing. Everywhere around me loomed glass, geologic in its permanence and grandeur: towers of glass, spires of glass, bluffs and fins and ravines of it, a million deceptive facets of sky glittering overhead.

I wandered along a courthouse facade as transparent as an Apple Store. In the revolving door squatted a white-throated sparrow (*Zonotrichia albicollis*), head striped in jaunty cream and black, who'd blundered into the window on his way north from the Gulf Coast. I approached and he fluttered against the door, left wing drooping. I backed off and watched him; he watched back, head cocked. The workday's first pedestrians stomped past, collars pulled up, oblivious to the tiny being struggling at their feet. A man burst out of the lobby, sending the door spinning, and the sparrow hopped along ahead of its sweeping pane.

Before long Prince arrived—a compact, competent-looking woman in a fluorescent raincoat. After greeting me, she took a moment to sweep the sparrow up in a butterfly net, slipped him into a paper bag, and laid the bag in a microwave-sized cardboard box that she carried under one arm. Although I'd feared that the sparrow would never fly again, Prince suspected he'd merely fractured his coracoid, a shoulder bone that a rehabber could heal with a wing wrap. "It was alert and moving—that's a good sign," she said.

Sparrow in tow, we walked through the city, canvassing storefronts and alleys where birds might be lying, stunned or dead. Elsewhere in Chicago, other volunteers searched their own neighborhoods. This was the Monitors' twentieth year in operation; Prince, a retired speech pathologist and avid birder, had been part of the group nearly since its inception.

Each morning during spring and fall migrations, its volunteers perused the street for victims and responded to reports that members of the public called in to a 24-hour hotline. Injured birds went to the Willowbrook Wildlife Center, a sanctuary outside the city, to receive treatment and eventually be released. Dead ones went to the city's Field Museum to enter its permanent collection. Most mornings, the dead outnumbered the living three to one.

Every year, the monitors collect around 7,000 birds, doubtless a tiny fraction of the unknowable number that die every year. Some days the work is constant: One recent October morning, the Monitors scooped up around 1,000 birds at McCormick Place, a convention center abutting Lake Michigan whose massive glass façade makes it a particularly egregious hotspot. Prince joked that the volunteers measured their busyness in Valium gulped. "People call and say, hey, is there some kind of disease outbreak going around?" she said wryly. "No, it's just architectural design."

Prince's phone rang: a bird reported to the hotline, in a neighborhood without a monitor. We got into her car and tore off, Prince weaving through traffic with a cabbie's reckless surety. When we arrived at the building—a preschool and hotel fronted by sheer glass—we found a female yellow-bellied sapsucker (*Sphyrapicus varius*), her eyes sunken and legs gone stiff.

"Every building has its own pathology," Prince had told me earlier. Here, it was easy to diagnose. Mirrored glass abutted a few scrawny trees outside the building, creating a faux, fatal forest: an optical illusion perfectly designed to slaughter birds. "You can see what a funhouse mirror this city is," Prince said as she stuffed the sapsucker into a bag—a plastic one.

The sapsucker's death was tragic both for the individual bird, and for all of avian kind. Since 1970, according to one large-scale synthesis of national bird surveys, U.S. bird populations have declined by close to 30 percent, a loss of nearly three billion animals. The culprits are many, but glass is among the most catastrophic, second only to cats as a direct killer. Certain bird species are unusually susceptible, according to a 2020 analysis, including wood thrushes (*Hylocichla mustelina*), yellowthroats (*Geothlypis trichas*), black-throated blue warblers (*Setophaga caerulescens*), and sapsuckers. That may be because these forest-dwelling migrants are accustomed to darting through tree canopy gaps at high speeds. For these and other vulnerable species, glass poses an existential threat.

Moreover, whereas cats or hawks often take out weaker or less wary animals, glass is an undiscerning predator, as apt to eliminate healthy migrants as sick ones. Our dead sapsucker was a hale breeding female who would have reared chicks this summer and likely for several to come. No longer. "What we've done here is killed one of the strongest members of her species," Prince said with a disgusted shake of her head. "We're incrementally taking away their future."

The history of bird-strike research in Chicago dates to 1978. That year, Dave Willard, an ornithologist at the city's Field Museum, received a tip from a coworker that birds were dying in droves at McCormick Place. He wandered over and found several casualties, among them a Virginia rail (*Rallus limicola*), a species then absent from the museum's collection. The birds slain at McCormick, Willard realized, were an untapped scientific resource—a near-endless font of data, just waiting to be harvested and analyzed.

Over time, Willard came to realize he'd also stumbled upon a catastrophe. Around 2,000 birds died at McCormick Place every year, in large measure because the brightly lit convention center lured nocturnal migrants into peril. Willard's documentation of the carnage eventually helped to inspire Lights Out Chicago, a citywide campaign that encourages building managers to extinguish or dim their lights during migration seasons; most years, more than 90 percent of Chicago's towers participate. Yet the harms of glass—a material, after all, that can't be turned on and off with the flick of a switch—have proved harder to address. "Even with lights out, there's a tremendous number of birds [being killed],"

Willard told me. "There's no change in our taste for glass as an architectural feature."

Although technically retired, Willard still works in a sort of emeritus capacity in the Field Museum, where I met him one afternoon. He led me into a labyrinth of cabinets and pulled open a tray packed with birds, most of which he'd collected and dressed in the early 1980s. I nearly gasped at the menagerie within, plumage undulled by four decades in a drawer: the cherry of cardinals (*Cardinalis cardinalis*), the lemon of meadowlarks (*Sturnella neglecta*), the ultramarine of an indigo bunting (*Passerina cyanea*). Each almost unbearably bright and beautiful; each gone too soon.

Willard reached into the tray and extracted the collection's most remarkable member: a brown creeper, an unassuming bird that makes its living plucking insects from bark. On April 17, 2008, he explained, volunteers had found this very creeper stunned on North Michigan Avenue. They'd taken it to the sanctuary, slipped a metal band around its ankle, and released it. A year later, precisely to the day, monitors picked up the exact same bird on the exact same street, this time dead. Now the bird lay in Willard's palm, a delicate testament to the gobsmacking fidelity of migratory birds, and to the inescapable doom wrought by the glass that blocks their way.

Today Annette Prince and her fellow volunteers annually deliver several thousand birds to the museum. Willard doesn't generally skin these specimens, as he once did; instead, he measures their weight and the length of their wings, bills, and feet, then feeds them to ravenous, flesh-devouring beetles to reduce them to skeletons. (The "beetle room" smells precisely as foul as you'd expect.) Once the insects have done their grisly work, Willard remands each set of bones into the kind of small beige box that, in another context, might contain a wedding ring. When I visited, boxes covered desks and tables in endless stacks, like the contents of a macabre and miniature Amazon warehouse.

As the museum's collection has grown, its birds have supported valuable research. In a 2019 study, Willard and his colleagues drew upon four decades of specimens to show that migratory birds are simultaneously losing weight as the climate changes—perhaps because it's advantageous to have a lighter body in a warmer world—and growing longer wings to compensate for their less robust physiques. He and others have also shown that species that call to each other in flight, like white-throated sparrows, are more likely to hit glass, possibly because flock members are attracted to light and then draw their comrades toward danger, like pilot whales beaching en masse. Windows are at once revelatory and tragic, offering insight into the lives of birds even as they obliterate them.

Yet Willard was that rare scientist who craved less data, not more. He opened a box to reveal the yellowed bones of a golden-crowned kinglet, so light and fragile it seemed they'd crumble to dust if I blew on them. He smiled sadly. "We'd trade everything we've learned," he said, "not to have them dead on the sidewalk."

For as long as buildings have sported glass, birds have likely collided with it. In an 1832 ornithology textbook, the naturalist Thomas Nuttall related the tale of a young male hawk who, while "descending furiously and blindly upon its quarry," smashed through a greenhouse. Miraculously, the hawk was "little stunned," though his "wing-feathers were much torn."

In Nuttall's day, glass was comparatively rare: windows tended to be small and set within brick or granite. Today it's everywhere—particularly in Chicago, longtime home of the mid-century architect Ludwig Mies van der Rohe, whose preference for vast glass facades still influences the city's aesthetic. Van der Rohe's purpose, he once said, was to fuse nature, humans, and structures in a "higher unity." The virtue of glass was that it connected indoor spaces with outdoor ones. The irony is awful: We prize a material that kills birds because it makes us feel closer to nature.

Yet even a perilous building can be made safer. One day, I took a self-guided tour of the Chicago area's bird-friendly architecture. I started in Evanston, home of Northwestern University, which had retrofitted a couple of particularly deadly buildings in response to monitors' data. Most problematic was the Kellogg Global Hub, a business-school headquarters as colossal and vitreous as an airport terminal. In 2018, Northwestern had coated part of Kellogg's facade with a translucent, dot-patterned film designed to make the building visible to birds. The dots, which were so faint that human passersby were unlikely to notice them, were spaced about as far apart as the width of my palm. Any wider than that, and birds would attempt to fly between the dots, as they flit through dense twigs and leaves. (A single hawk decal on a big pane? Essentially useless.)

The film seemed to be working: Collisions at the Kellogg Hub had declined by half, and for twenty minutes I watched red-winged blackbirds (*Agelaius phoeniceus*) alight easily on its railings and roof. Even better, at the nearby Frances Searle Building, whose windows the university had covered with faint horizontal stripes, bird deaths had virtually ceased. Yet the projects had been neither cheap nor perfect. The Searle Building alone had cost the university $250,000 to retrofit, and large swaths of the Kellogg Center remained unprotected. Retrofitting existing buildings is crucial, no doubt; Chicago isn't about to dismantle its existing skyline for the sake of birds. Still, "the best solutions are the ones that are designed into the building from the beginning," Claire Halpin, a landscape architect who sat on the board of the Chicago Ornithological Society until her recent death, told me later.

Few architectural firms do that better than Halpin's former employer, Studio Gang, the firm behind some of Chicago's bird-friendliest mega-structures. I visited two of them, starting with the Aqua Tower, an eighty-two-story monolith frilled with curvaceous balconies, as though the building had sprouted shelf fungi. The terraces lent the tower "visual noise," warning birds that this otherwise reflective structure was in fact a solid object. The studio had applied similar principles at Solstice on the Park, an apartment complex whose glass panels were angled toward the ground and thus reflected earth rather than sky. The lobby's windows were also subtly covered with dashes—a material, known as "fritted" glass, with markings printed on the pane rather than added retroactively. Both buildings, I noticed, incorporated enormous expanses of glass, yet they possessed a visibility that other Chicago towers lacked.

"It's a mistaken view that bird-friendly architecture means you have to build this opaque thing that nobody can see in or out of," Halpin said. "Transparency does not have to be in conflict with birds."

What's more, avian safety doesn't always require structural overhauls. During migration season, the Federal Bureau of Investigation swaddles its Chicago headquarters in fine black mesh, off of which birds harmlessly bounce. At the Blue Cross Blue Shield tower, birds used to die often while trying to reach a potted ficus stationed invitingly in the lobby. At Prince's suggestion, the building's managers moved the plant away from the window, and the collisions virtually ceased.

Not everyone was so receptive. During our rounds, Prince and I passed the Bank of America Tower, where three trees beckoned to birds from behind glass so clear it was surprising more humans didn't hit it face-first. Prince had begged the building's sustainability officer to put up a patterned film, or move the trees, or even hang a temporary banner during migration season—to do literally *anything*. "The answer," she told me, "was 'no, no, and no.'"

Fortunately, Chicago's bird advocates are also attempting to influence policy and compel widespread change. In 2021, Illinois began to require new state-owned buildings to incorporate netting, screens, shutters, and other bird-friendly features, one of the country's first such laws. Even more promising, in 2020 Chicago passed an ordinance mandating that new buildings limit their use of transparent and reflective glass, use patterned glass in high-risk areas, and reduce the night-time interior lighting that lures birds to their death. The city's Department of Planning and Development expects to implement the ordinance by the end of this year.

This progress hasn't been greeted with pleasure by all developers, some of whom fear that patterned glass will jack up construction costs and deter retailers from renting space. As one put it to *Landscape Architecture Magazine*, "There is a real big bird that this ordinance is going to kill: the biggest bird in town, the goose that laid the golden egg in real estate."

Yet the notion that protecting birds harms business is largely a myth. For one thing, because nearly all bird collisions happen in the lowest hundred feet, architects don't need to treat entire high-rises with bird-friendly glass; instead, the bottom seven stories generally suffice. For another, glass represents such a minuscule portion of construction costs that spending a little extra has virtually no effect on a building's bottom line. In a 2022 report, Daniel Klem, an ornithologist at Muhlenberg College who's studied window collisions since the 1970s, found that bird-safe glass adds less than four-tenths of a percent to the cost of a typical building. (An $8 million office tower, for example, would only pay an additional $30,000 or so.) Moreover, Klem argued, as pro-bird laws in Illinois, Minnesota, Maryland, and other states drive up demand for bird-friendly glass, glass manufacturers will inevitably produce more of it and lower their prices.

"Right now, the majority of developers and architects don't have this issue on their radar, but many are changing," Klem told me. Years ago, he said, a magazine had branded him the "Rodney Dangerfield of ornithology," referencing the comedian whose trademark joke was that he never got the respect he deserved. Within the past decade, however, respect for bird collisions has arrived, if belatedly. "Members of these key constituencies are joining the cause of saving more lives from windows," Klem said. "These are innocent creatures that need our help."

At the end of our patrol, Prince took me to a parking lot where a car, one door emblazoned with a Bird Collision Monitors panel, had been left unlocked—with windows cracked and interior pleasantly cool—so that surveyors could drop off their bounty. In the backseat, a cardboard box brimmed with dead birds bound for the Field Museum: an eastern towhee (*Pipilo erythrophthalmus*), a yellow-rumped warbler (*Setophaga coronata*), a veritable flock of white-throated sparrows.

Happily, the hatchback was also full of paper bags that rustled faintly as their captives stirred—the live animals that a volunteer would soon drive out to the sanctuary, and that might yet continue interrupted journeys.

Prince inspected the bags' sharpied labels. Woodcock. Virginia rail. Red-headed woodpecker (*Melanerpes erythrocephalus*). "That one's probably a mislabeled sapsucker," she said. She reached in and removed a gorgeous creature—wings and breast feathered in dapper black and white, vivid maroon from the shoulders up— sure enough, a red-headed woodpecker. He crankily clamped his bill around Prince's thumb.

"You are so beautiful. Oh, you're so pretty," Prince cooed. "You're gonna go out to a nice forest preserve where you can be away from these scary buildings." She turned to me, full of wonder. "Right in the middle of this dense urban area, we've got these exotic birds that people would be thrilled to see in the woods or a marsh."

Like glass itself, bird collisions are often invisible. Songbirds are tiny and delicate; thousands of loafer-clad feet step over them, and rats and crows scavenge them. Clever gulls have even learned to chase songbirds into buildings and snatch them as they fall limp—resident birds exploiting the naïveté of migrants. Cities devour birds and grind them to pulp, a crisis both omnipresent and mostly unremarked.

The beauty of monitoring is that it reveals this hidden catastrophe: that it alerts a great many more people to the millions of small lives that wing through cities, and even converts them into advocates. At one point, as Prince and I drove around Chicago, she took a call from a doorman on Wacker Street. We raced over and found a red-breasted nuthatch (*Sitta canadensis*) huddled against a planter, so dazed that Prince could pick her up without a net. "Did you find it?" the doorman called, his voice anxious. His nametag read Fernando. Prince showed him the nuthatch in her palm and his eyes widened. "Cool, cool," Fernando said, with evident relief.

Prince unfolded a paper bag, slipped the nuthatch inside, and stowed it in the trunk. Fernando watched, riveted. "Hey, could I get a couple of those bags?" he asked. "For next time?" Prince was delighted to oblige.

DOUGLAS FOX

Journey Under the Ice

FROM *Science News*

THE COASTAL PLAIN of the Kamb Ice Stream, a West Antarctic glacier, hardly seems like a coast at all. Stand in this place, 800 kilometers from the South Pole, and you see nothing but flat ice extending in every direction. The ice is some 700 meters thick and stretches for hundreds of kilometers off the coastline, floating on the water. On clear summer days, the ice reflects the sunlight with such ferocity that it inflicts sunburn in the insides of your nostrils. It might seem hard to believe, but hidden beneath this ice is a muddy tidal marsh, where a burbling river wends its way into the ocean.

Until recently, no human had ever glimpsed that secret landscape. Scientists had merely inferred its existence from the faint reflections of radar and seismic waves. But in the closing days of 2021, a team of scientists from New Zealand melted a narrow hole through the glacier's ice and lowered in a camera. They had hoped that their hole would intersect with the river, which they believed had melted a channel up into the ice—a vast water-filled cavity, nearly tall enough to hold the Empire State Building and half as long as Manhattan. On December 29, Craig Stevens finally got his first look inside. It is a moment that he will always remember.

Stevens is a physical oceanographer with New Zealand's National Institute of Water and Atmospheric Research in Wellington. He spent ninety anxious minutes that day in Antarctica with his head buried ostrich-style under a thick down jacket to block the sunlight that would otherwise obscure his computer monitor. There, he watched live video from the camera as it descended into the hole. Icy circular walls scrolled past, reminiscent of a cosmic wormhole. Suddenly, at a depth of 502 meters, the walls widened out.

Stevens shouted for a colleague to halt the winch lowering the camera. He stared at the screen as the camera rotated idly on its cable. Its floodlights raked across a ceiling of glacial ice—a startling sight—scalloped into delicate crests and waves. It resembled the dreamy undulations that might take millennia to form in a limestone cavern.

"The interior of a cathedral," says Stevens. A cathedral not only in beauty, but also in size. As the winch restarted, the camera journeyed downward for another half hour, through 242 meters of sunless water. Bits of reflective silt stirred up by currents streamed back down like snowflakes through the black void.

Stevens and his colleagues spent the next two weeks lowering instruments into the void. Their observations revealed that this coastal river has melted a massive, steep-walled cavern cutting as far as 350 meters up into the overlying ice. The cavern extends for at least 10 kilometers and appears to be boring inland, farther upstream, into the ice sheet with each passing year.

This cavity offers researchers a window into the network of subglacial rivers and lakes that extends hundreds of kilometers inland in this part of West Antarctica. It's an otherworldly environment that humans have barely explored and is laden with evidence of Antarctica's warm, distant past, when it was still inhabited by a few stunted trees.

One of the biggest surprises came as the camera reached bottom that day. Stevens gazed in disbelief as dozens of orange blurs swam and darted on his monitor—evidence that this place, roughly 500 kilometers from the open, sunlit ocean, is nonetheless bustling with marine animals.

Seeing them was "just complete shock," says Huw Horgan, a glaciologist formerly at the Victoria University of Wellington who led the drilling expedition.

Horgan, who recently moved to ETH Zurich, wants to know how much water is flowing through the cavern and how its growth will impact the Kamb Ice Stream over time. Kamb is unlikely to fall apart anytime soon; this part of West Antarctica is not immediately threatened by climate change. But the cavern might still offer clues to how subglacial water could affect more vulnerable glaciers.

Mapping the unknown

Scientists have long surmised that a veneer of liquid water sits beneath much of the ice sheet covering Antarctica. This water forms as the bottom of the ice slowly melts, several penny-thicknesses per year, due to heat seeping from the Earth's interior. In 2007, Helen Amanda Fricker, a glaciologist at the Scripps Institution of Oceanography in La Jolla, Calif., reported evidence that this water pools into large lakes beneath the ice and can flood quickly from one lake to another.

Fricker was looking at data from NASA's Ice, Cloud and Land Elevation Satellite, or ICESat, which measures the height of the ice surface by reflecting a laser off of it. The surface at several spots in West Antarctica seemed to bob up and down, rising and falling by as much as nine meters over a couple of years. She interpreted these active spots as subglacial lakes. As they filled and then spilled out their water, the overlying ice rose and fell. Fricker's team and several others eventually found over 350 of these lakes scattered around Antarctica, including a couple dozen beneath Kamb and its neighboring glacier, the Whillans Ice Stream.

The lakes provoked great interest because they were expected to harbor life and might provide insights about what sorts of organisms could survive on other worlds—deep within the ice-covered moons of Jupiter and Saturn, for instance. The layers of sediment in Antarctica's lakes might also offer glimpses into the continent's ancient climate, ecosystems and ice cover. Teams funded by Russia, the United Kingdom and the United States attempted to drill into subglacial lakes. In 2013, the U.S.-led team succeeded, melting through 800 meters of ice and tapping into a reservoir called Subglacial Lake Whillans. It was teeming with microbes, 130,000 cells per milliliter of lake water.

Horgan helped map Lake Whillans before drilling began. But by the time the lake was breached, he and others were becoming intrigued with another facet of the subglacial landscape—the rivers thought to carry water from one lake to another, and eventually to the ocean.

Finding these hidden rivers requires complicated guesswork. Their flow paths are influenced not only by the subglacial topography, but also by differences in the thickness of the overlying ice. Water moves from places where the ice is thick (and the pressure high) to places where it is thinner (and the pressure lower)— meaning that rivers can sometimes run uphill.

By 2015, scientists had mapped the likely paths of several dozen subglacial rivers. But drilling into them still seemed farfetched. The rivers are narrow targets and their exact locations often uncertain. But around that time, Horgan got a lucky break.

While examining a satellite photo of the Kamb Ice Stream, he noticed a wrinkle in the pixelated tapestry of the image. The wrinkle resembled a long, shallow trough in the surface of the ice, as if the ice had sagged from melting beneath. The trough sat several kilometers from the hypothetical path of one subglacial river. Horgan believed that it marked the spot where that river flowed over the coastal plain and spilled into the ice-covered sea.

In 2016, while visiting the area for an unrelated research project, Horgan and his companions detoured briefly to the surface trough to take radar measurements. Sure enough, they found a void under the ice, filled with liquid water. Horgan began making plans to study it more closely. He would return twice in the next few years, once to map the river in detail and a second time to drill into it. What he found greatly exceeded his expectations.

A river runs through

Horgan and graduate student Arran Whiteford of the Victoria University of Wellington visited the lower Kamb Ice Stream to map the river in December 2019.

After weeks on the Antarctic ice sheet, they'd grown accustomed to its monotonous flat landscape, their perception sensitized to even tiny ups and downs. In this context, the surface trough "looked like this massive chasm," Whiteford says, "like an amphitheater"— even though it slanted no more dramatically than a rolling cornfield in Iowa.

It was a week of scientific drudgery, towing the ice-penetrating radar behind a snowmobile along a series of straight, parallel lines that crisscrossed the trough to map the shape of the river channel under the ice.

Horgan and Whiteford worked up to twelve hours per day, occasionally trading positions. One person drove the snowmobile, straining his thumb on the throttle to maintain a constant 8 kilometers per hour. Two sleds hissed along behind. One held a transmitter that fired radar waves into the glacier below; the other held an antenna that received the signal reflected back off the bottom of the ice. The second person rode on the sled with the antenna, his eyes on a bouncing laptop screen making sure that the radar was functioning.

Each evening they huddled in their tent, reviewing their radar traces. The river channel appeared far more dramatic than the gentle dip atop the ice suggested. Below their boots sat a vast water-filled cavern with steep sides like a train tunnel, 200 meters to a kilometer wide and cutting as much as 50 percent of the way up through the glacier. The more they looked, the more it resembled a river. "It kind of meanders downstream," Whiteford says.

All told, Whiteford made two weeklong visits to the trough, snowmobiling over from another camp 50 kilometers away. The first time he was accompanied by Horgan, and the second time by another graduate student, Martin Forbes.

After returning home to New Zealand in January 2020, Whiteford examined a series of old satellite images. They showed that the surface trough—and hence, the cavern—had begun forming at least thirty-five years before, starting with a blip at the very mouth of the river, where it ran into the ocean. That blip had gradually lengthened, reaching progressively farther inland, or upstream. Whiteford and Horgan reported the observations in late 2022 in the *Journal of Geophysical Research: Earth Surface*—along with their theory about how the cavern formed.

In other parts of Antarctica where the ice sheet protrudes off the coastline, scientists have found that the ice's underside is often insulated from the ocean heat by a buoyant layer of colder, fresher meltwater. That protective layer is sometimes only a couple of meters thick. But Horgan and Whiteford suspect that the turbulence of the subglacial river flowing into the ocean stirs up that protective layer, causing seawater—a few tenths of a degree warmer than the subglacial water—to swirl up into contact with the ice. This causes an area of concentrated melt right at the river's mouth, creating a small cavity where warm seawater can intrude further.

In this way, says Horgan, the focal point of melting is "stepping back over time." And the cavern gradually burrows farther upstream into the ice.

Whiteford used a different set of satellite measurements—which measured the rate at which the ice's surface sank over time—to determine how quickly the ice was melting in the cavern below. Based on this, he estimated that in the upstream end of the cavern, the ice (currently 350 to 500 meters thick over the channel) was melting and thinning 35 meters per year. That's an astronomical rate. It's 135 times what has been measured 50 kilometers southwest of the cavern, where the ice floats on the ocean. The water temperature is probably similar at both locations. But the turbulence caused by the river transfers the water's heat far more efficiently into the ice.

Horgan thinks that the cavern at Kamb also owes its dramatic height to another factor. Glaciers in this part of West Antarctica generally flow several hundred meters per year. So the melt caused by a flowing river beneath, over years or decades, would normally be spread out over a long swath of ice. This would erode a shallow channel rather than a deep cleft. But Kamb is an oddball. Around 150 years ago, it stopped moving almost entirely due to the cyclical interplay of melting and freezing at its base. It now creeps forward only about 10 meters per year. The melting is thus concentrated, year after year, in almost the same spot.

Back in 2020, all of this was still conjecture. But if Horgan and his colleagues could return, drill into the cavern and lower instruments into it, they could confirm how it formed. By studying the water, sediment and microbes flowing out of it, they could also learn a lot about Antarctica's vast subglacial landscape.

The West Antarctic Ice Sheet covers an area three times the size of the Colorado River drainage basin, which sprawls across Arizona, Utah, Colorado and parts of four other states. To date, humans have observed only a tiny swath of this underworld, smaller than a basketball court—represented by several dozen narrow boreholes scattered across the region, where scientists have grabbed a bit of mud from the bottom or sometimes lowered in a camera.

Horgan was eager to explore more. With New Zealand already

melting boreholes through ice floating on the ocean, drilling into this coastal river seemed like a natural next step.

Looking to the lakes

On December 4, 2021, a pair of caterpillar-tracked PistenBullys arrived at the place where Horgan and Whiteford had visited two years before. The tractors had traveled for sixteen days from New Zealand's Scott Base on the edge of the continent, growling across a thousand kilometers of floating ice as they towed a convoy of sleds packed with 90 metric tons of food, fuel and scientific gear. The convoy lumbered around to the upstream end of the valley and stopped.

Workers erected a tent the size of a small aircraft hangar, and inside it, assembled a series of water heaters, pumps and a kilometer of hose—a machine called a hot water drill. Using shovels and a small mechanized scooper, they dumped 54 tons of snow into a tank and melted it. The workers then jetted that hot water through the hose, using it to melt a narrow hole, no wider than a dinner plate, through 500 meters of ice—and down through the domed ceiling of the cavern.

The sight of animals inside the cavern generated instant excitement among Horgan, Stevens and the other people at camp. But those first images were blurry, leaving people unsure of what the orange, bumblebee-sized critters actually were.

Workers next lowered an instrument down the borehole to measure the water temperature and salinity inside the cavern. They found the top 50 meters of water colder and fresher than what lay below—confirming that seawater was flowing in along the bottom and a more buoyant mixture of saltwater and freshwater was flowing out along the top. The cavern, says Stevens, "is operating quite like an estuary."

But those measurements also presented a mystery: The water in the top of the cavern was only about 1 percent less salty than the seawater in its bottom, suggesting that the amount of freshwater flowing in through the river was "quite small," says Stevens. It's akin to a shallow creek that a young kid might splash around in. He and Horgan doubted that the turbulence caused by this small

flow, even over thirty-five years, could melt the entire cavern—
roughly a cubic kilometer of ice.

A likely answer came from a set of samples collected from the
floor of the cavern. Gavin Dunbar, a sedimentologist at the Victoria University of Wellington, lowered a hollow plastic cylinder
down the hole in hopes of retrieving a core. As he and graduate
student Linda Balfoort hoisted the cylinder back up, they found
it streaked and filled with chocolaty mud—a strange sight in this
world of pure white, where not a speck of rock or dirt can be seen
for hundreds of kilometers.

As Dunbar and Balfoort X-rayed and analyzed the cores months
later, back in New Zealand, their peculiarities became obvious:
They were unlike anything that Dunbar had ever encountered in
this part of the world.

Every core that Dunbar had ever seen from the seafloors near
this part of Antarctica consisted of a chaotic jumble of sand, silt
and gravel—a material called diamict, formed as the ice sheet advances and retreats over the seafloor, plowing and mixing it like a
rototiller. But in these cores, Dunbar and Balfoort saw distinct layers. Bands of coarse, gravely material were interspersed with layers
of fine, silty mud.

That alternating pattern resembled samples from steep seafloor
canyons off the coast of New Zealand, where earthquakes sometimes trigger underwater landslides that sweep for many kilometers
downhill. Each flood deposits a single layer of chunky material.

Dunbar believes that something similar happened under the
Kamb Ice Stream, possibly in the last few decades. A series of fast-
moving torrents gushed through the river channel carrying big
gravelly chunks from somewhere upstream that later settled on the
cavern floor. "Each of these [coarse layers] represents minutes to
hours of sediment deposition" that occurred during a single flood,
he says. And the fine, silty layers would have been laid down over
years or decades in between the floods, when the river flowed lan-
guidly along.

These subglacial floods could explain how this small river
carved such a large cavern, Stevens says. Those floods could have
been 100 to 1,000 times as large as the flow rates that were measured during the 2021–22 field season.

No one knows when those events happened, but scientists using satellites to study subglacial lakes have spotted at least one candidate. In 2013, a lake 20 kilometers upstream from the cavern, called KT3, disgorged an estimated 60 million cubic meters of water—enough to fill 24,000 Olympic-sized swimming pools.

Scientists would love to know whether that flood actually passed through this cavern. "Connecting this upstream to the lake system would be extremely cool," says Matthew Siegfried, a glaciologist at the Colorado School of Mines in Golden, who coauthored one of the reports documenting the 2013 flood.

Studying the outflow of this river could also answer other questions about the subglacial landscape upstream. "The vast majority of our knowledge of subglacial lakes comes from surface observations from space," Siegfried says. But those satellite records, of ice bobbing up and down, permit only indirect estimates of how much water is flowing through. It's possible, for example, that a lot of water passes through the lakes even when the ice above isn't moving.

Scientists could also learn about the subglacial landscape by studying the sediment washed downstream. When Dunbar and his colleagues examined the coarse material from their cores, they found it full of microscopic fossils: glassy shells of marine diatoms, needly spicules of sea sponges, and notched and spiky pollen grains of southern beech trees. These fossils represent the remains of a warmer world, 15 million to 20 million years ago, when a few stands of stunted, shrubby trees still clung to parts of Antarctica. Back then, the West Antarctic basin held a sea rather than an ice sheet, and this detritus settled on its muddy bottom. These old marine deposits underlie much of the West Antarctic Ice Sheet, and the few boreholes drilled so far suggest that the mix of fossils differs from one place to another. Those mixes could provide clues to how the flow of rivers changes over time.

To uncover the nuance of what's happening in the cavern "is mind-blowingly cool," says Christina Hulbe, a glaciologist at the University of Otago in Dunedin, New Zealand, who has studied this region of Antarctica for nearly thirty years. "That's the outlet for a massively big river system, if you think about it."

By studying the water, scientists could estimate the amount of organic carbon and other nutrients flowing out of the river into the ice-covered ocean. The landscape beneath the ice sheet appears to be rich in nutrients that might sustain oases of life in an otherwise famished biological desert.

An oasis of life

Even as the cavern penetrates farther into the Kamb Ice Stream, it does not necessarily threaten the glacier's stability. This part of the West Antarctic coastline is not considered vulnerable, because its shallow bed shields it from the deep, warm ocean currents that are causing rapid ice loss in other regions. But subglacial rivers pour out at many other points along the coastline, including some—like Thwaites Glacier, roughly 1,100 kilometers northeast of Kamb—where the ice is retreating rapidly.

Thwaites and nearby glaciers have collectively shed over 2,000 cubic kilometers of ice since 1992. They could eventually raise global sea levels by 2.3 meters if they collapse. Remote sensing studies have documented over a dozen low, squat shield volcanoes beneath this part of the ice sheet. The elevated geothermal heat flow, even from inactive volcanoes, is thought to cause high levels of melting under the ice sheet. That melting produces large amounts of subglacial water, which could render these glaciers even more vulnerable to human-caused climate change.

Horgan believes that what scientists learn at Kamb could improve our understanding of how subglacial rivers impact those other, rapidly changing coastlines of Antarctica.

But the most evocative discovery made at Kamb—in purely human terms—may be the blurry, orangish animals seen swarming near the bottom of the cavern. Stevens captured some clearer images a few days later and tentatively identified them as shrimp-like marine crustaceans called amphipods. To see so many of them here, Stevens says, "we really hadn't expected that."

Microbes like those previously found under the ice sheet in Subglacial Lake Whillans are known to eke out a living even in harsh conditions. But animals are a different matter. The deepest seafloors on Earth sit only 10 or 11 kilometers from sunlight, and animal life in those places is generally scarce. But the animals in the cavern are thriving 500 kilometers from the nearest daylight, cut off from the photosynthesis that fuels most life on Earth.

The amphipods and their supporting ecosystem must be subsisting on some other food source. But what? Observations in the Kamb ice cavern, combined with those at two other remote bore-holes drilled in recent years, offer some tantalizing hints.

In 2015, researchers pierced the ice at another site 250 kilome-ters from the cavern, where the Whillans Ice Stream lifts off its bed and floats. In that location, a thin sliver of seawater, just 10 meters deep, sits beneath 760 meters of ice. A remotely operated vehicle, or ROV, sent down the hole captured images of fish and amphipods.

John Priscu, a microbial ecologist at Montana State University in Bozeman who was involved in the drilling at the site, believes that the glacier itself is sustaining this ecosystem. The bottom 10 meters of ice is packed with mud that had frozen onto the belly of the gla-cier many kilometers upstream. The mud had been dragged to its present location as the glacier oozed forward, 400 meters per year. As the ROV navigated about, bits of that muddy debris constantly rained down, released as the ice's underside slowly melted. That de-bris is rich in organic matter—the rotting remains of diatoms and other phytoplankton that sank to the bottom millions of years ago when the world was warmer.

"Those amphipods are swarming to the particulate matter," Priscu says. "They're sensing the organic matter falling out of that basal ice." Or perhaps they may be eating the bacteria that live on those organics.

Because the Kamb Ice Stream is barely moving, the supply of dirty ice moving toward the sea is small. But the river flowing into the ice cavern may deliver the same subglacial nutrients that are found in dirty ice. After all, the water's journey through a series of subglacial lakes down to the river's mouth may take years or decades. Throughout that time, the river absorbs nutrients from the organic-rich subglacial sediments.

Indeed, when scientists drilled into Subglacial Lake Whillans in 2013, they found its water honey-colored—chock-full of life-sustaining iron, ammonium and organics. "What these lakes are pumping out may be a concentrated source of nutrients" for eco-systems along the dark coastline, says Trista Vick-Majors, a micro-bial ecologist at Michigan Technological University in Houghton who was involved in the drilling at Lake Whillans. She has esti-mated that the subglacial rivers flowing out from under Kamb and its neighboring glaciers may deliver 56,000 tons of organic carbon and other nutrients to this section of the coastline every year.

More recently, in December 2019, a team from New Zealand led by Horgan and Hulbe drilled through the ice just 50 kilometers from the Kamb cavern, in a place where the Kamb Ice Stream floats on the ocean. There's no dirty ice there and no nearby river outlets. The area resembled a famished seafloor desert; it was populated by single-celled microbes with little to eat, and few signs of animals were seen—only a few burrowing traces on the muddy bottom. Priscu sees this location as an exception that proves the point: Subglacial nutrients are the crucial energy source in this dark world under the floating ice, whether they are dragged forward on the undersides of glaciers or spilled out through subglacial rivers.

The mud and water samples collected from the Kamb ice cavern may provide a new opportunity to test that theory. Craig Cary, a microbial ecologist at the University of Waikato in New Zealand, is analyzing DNA from those samples. He hopes to determine whether the microbes in the cavern belong to taxonomic groups that are known to subsist on ammonium, methane, hydrogen or other sources of chemical energy that originate from the subglacial sediments. That might reveal whether such sources support enough microbial growth to feed the animals observed there.

The team also needs to measure the flow rate of the subglacial river that spills into the cavern, since that determines the nutrient supply. Stevens continues to monitor this thanks to a set of instruments left behind in the cavern.

As people were packing up camp on January 11, 2022, workers pumped more hot water into the borehole, widening it to more than 35 centimeters—and creating a dangerous pitfall. Stevens and his colleagues donned climbing harnesses, clipped into safety ropes and approached the hole one last time. They lowered a series of cylinders the size of caulking guns down the hole. These devices continue to measure the temperature, salinity and water currents inside the cavern, sending the data 500 meters up a cable to a transmitter that beams it home via satellite once a day. That data will reveal how the river's flow changes over time. With luck, the instruments might even detect a subglacial flood gushing through.

"That would just be outstanding," Horgan says. For many years, he had to content himself with seeing these rivers and lakes dimly, through the outlines of water on radar and satellite images. This is "one of the first times we've got to stand at a river mouth and observe it."

Five Hours from Someplace Beautiful

FROM *Oxford American*

Gray Fox
(Urocyon Cinereoargenteus)

One early morning, a few months after my daughter was born, I ducked out to my backyard to turn on the sprinkler and found a gray fox balancing on our wood fence. As agile as any house cat, it prowled the fence top, maybe thirsty for the wading pool that had previously slumped in the shade of our desert willow. I threw the door behind me closed—an instinct—and the fox paused to look me in the eye. It stood straight and still, its flared, black-streaked tail raised like a flag.

I'd heard the hoarse yelp of a fox once before: sunset, walking the dirt easement that extended beyond the end of the road behind our subdivision. It was total shit land out there. A chunk of it was irrigated for grazing a herd of maybe a dozen cattle, there were high tension powerlines, a submerged gas pipeline, and some lots slated to be fracked and drilled, but mostly it was mesquite, bunchgrass, and plastic bags.

In that feral space, the sound of a wild animal had felt fitting, but perched on my fence, the fox became an uncanny reminder of the fragile claims we lay on the land we occupy.

Barn Owl
(Tyto Alba)

Everyone in Midland knows something beautiful is just a five-hour drive away. Bumper stickers from the Texas Hill Country, Big Bend National Park, Carlsbad Caverns, and the mountains of New Mexico and Colorado tattoo our pickups and SUVs. But the lights on the local prospect are somewhat dimmer. In a community built upon extraction, the value of its land is foremost a calculation of the barrels it can provide. Aesthetic pleasures lie elsewhere.

Promotional photos for the city often feature the downtown skyline rising above duck ponds and verdant slopes of golf courses—belying the truth of an environment that receives less than fifteen inches of precipitation in an average year. Yet, our very livelihood depends on expanses of space where so many of us spend our working days digging through dust.

Our home was built in 2014, following the height of the oil boom that drove a severe housing shortage. Workers poured into town, lured by lucrative jobs in the surrounding Permian Basin. Rents were higher in our city of about 130,000 people than in Dallas or Austin. Man camps, where temporary oilfield workers are housed, were full. Motels were full. RV parks were full. People were moving here and living in the cabs of their pick-up trucks while making thousands of dollars a week.

Housing developments like ours have popped up in waves along the verges of the city: an HOA, a community pool, a half dozen preset floorplans, your choice of brick façade. They're raised in a rush by corporate builders, sprawling out into former ranchland from a once compact city center to fill the nagging demand for housing.

Like most of the new middle-class neighborhoods of Midland, ours half-heartedly aspires to the slick charm of the suburban American dreamscape. Our front yards are made of gray river rock or squares of sod that never quite take to the soil. Pumpjacks teeter in quarter-acre lots set between cul-de-sacs. There always seems to be not quite enough trees and a few more trucks than anyone could need.

Our builder told us to expect some wildlife to appear our first year—scared out of hiding by the scrape of bulldozers and graders.

We found scorpions on the bathroom tile. Our neighbors were freaked by a rattler curled into the corner of their garage. One afternoon, my wife spotted a barn owl roosting in the just-framed eaves of a two-story across the street. It felt as if the moon itself had been displaced. We stood at the bottom of the driveway, watching until it yawned its broad wings and circled toward the seemingly empty fields that lay beyond the subdivision.

The next Saturday morning, I went walking in the direction of the owl, past construction sites where roofers and masons were already blasting corridos from their boomboxes. I was looking for some sense of place—an assurance, however vague, that my life is connected to the natural world and the history of the people in it. I had an impression that neighborhoods like ours are not *place* itself, but a palimpsest upon it.

To find the *here* here, I had to walk past the end of the road.

Tumbleweed
(Salsola Tragus)

Following the owl beyond the subdivision, on the first of many walks, I found myself in an ill-defined space: ugly, post-agricultural, metamorphic. Not the oilfield proper, and not exactly town—not yet, anyway. A catchment basin, a transformer station, then a patch of about two square miles of scrub.

British environmentalist Marion Shoard coined the term "edge-lands" to describe "the latest version of the interfacial rim that has always separated settlements from the countryside." Her edgelands seem to be places of neglect—where the guiding hand of the planner has vacated the premises, leaving battered nature and the people who move through it to their own devices. But here, where hunger for land and its resources is bottomless, any such *edge* is one that only pushes ever outward.

This territory I've come to love deserves a more honest name: shit land. Abused but resilient, transformed but still wild. You can find it everywhere in America, and it deserves our consideration.

In West Texas, this is where the plastic shopping sack becomes unlikely kin to the tumbleweed and the dragonfly—each animated by a similar wandering spirit. Everything is steered by the wind across

our half desert, and these three are driven—however circuitously—
toward the points of yucca, barbed wire, and mesquite.

On that first walk, I saw each of them congregating along a fence
that ran parallel with the roadway. A few days before, a Haboob,
a huge rolling dust storm, brought brown skies and left grit between my teeth. It'd also knocked loose stampedes of tumbleweeds.
They're an icon of the west, but their introduction to the United
States can be drawn to contaminated shipments of Russian flax to
South Dakota in the 1870s. Spread by railway and wind, they quickly
became emblematic of frontier solitude. They're hawked as rustic
tchotchkes on Etsy.

The shrubs grow three feet tall through the summer, covered
in fleshy green needles. As autumn approaches, green tinges red
and the needles curl up from the dirt in thick mats, revealing hundreds of thousands of seeds the parent plant will broadcast as it
dies, breaks away, and rolls across the plain.

As the morning began to warm, I found dragonflies resting in
the sun on strands of barbed wire. The green darner—large, lime
and electric blue—makes a multigenerational migration of up to
nine hundred miles, from Mexico to Canada. Though the patterns
of their seasonal movements are not well understood, it's believed
that they are able to read the prevailing winds of a given day and
lift themselves into currents that point them toward their seasonal
destinations.

The same fence that harbored darners and tumbleweeds also
collected plastic bags. Blown from parking lots and truck beds and
hooked on the fence points, they tattered in the breeze like Tibetan
prayer flags, broadcasting over the land the sutras of our twinned
local deities: commerce and petroleum.

Honey Mesquite
(Prosopis Glandulosa)

Alternately rutted and flattened by a procession of frack sand
haulers and bulldozers, the lane of parched red and yellow silt
where I walked for the better part of the last decade cut through
dilapidated pasture. Neon airplane bottles of 99 schnapps were
crunched into the ground from where they'd been tossed out the
window of one of the countless white company pickups—whether

in celebration or to ease the way through another shift was impossible to tell. Styrofoam confetti and bridal trains of Bubble Wrap littered the way toward a cattle guard on the horizon.

Walking one afternoon in early autumn beneath the buzz of the powerlines, I spotted a pair of Harris's hawks hunting together—one swooping into the brush while the other waited on a rung of the powerline tower for whatever prey might be flushed into the breaks. Black-tailed jackrabbits—long ears pricked, testing the air—stood on their hind legs before bounding ten or more feet at a leap, toward a ridge of mesquite that ran along one side of the road.

This was a windbreak, or the remains of one, planted to protect livestock and their feed from desiccating winds. Only a few decades ago, much of Midland County was devoted to cattle, but the pressure of the oil industry has left only remnants of ranching today.

Honey mesquite—a tangled shrub of thorns, feathery leaves, and slender bean pods—is the most conspicuous presence in our southern stretch of the Llano Estacado—the tableland that tilts up 250 miles from here to the border of Oklahoma. Prior to colonial settlement, mesquite was at home in canyons and draws—sloping courses of ephemeral streams—and its smoky-sweet beans were prized forage for the Comanche and Apache. But eradication of controls on its growth (the millions of bison killed to thwart native peoples, shortgrass species grazed bare) led to its invasive spread throughout former prairie. A bane to cattlemen, its deep roots sucked up precious moisture, and the pods sickened animals. The marketing genius that eventually turned a nuisance into potato chip flavoring didn't start until the 1950s.

Where it has been irrigated, mesquite can grow into small trees, collecting dunes at their feet. Their shade creates a niche for smaller shrubs, flowers, and vines like the buffalo gourd, whose triangular foliage and golden, baseball-sized fruits snake down through the jumble. It, too, has been naturalized here. A "camp follower," it likely migrated north in the wake of migrations ten thousand years ago.

Legacy upon legacy, the human hand has shaped this land to meet human needs, and wildness has fitted itself accordingly. I watched the mesquite where the Harris's hawks made their nest. In the hump of dirt and underbrush beneath, the jackrabbits hid from the hawks and from me.

Silverleaf Nightshade
(Solanum Elaeagnifolium)

Like the mesquite and buffalo gourd, many of the plants I've encountered on my walks are best classified as *ruderal*—those species
that, by dint of their prolificacy, speedy growth, and meager needs,
are capable of pioneering disturbed soils. But it's in the spirit of
shit land to call them by their common name: weeds.

Silverleaf nightshade is my favorite.

Late spring through autumn, its dry fingers of foliage prop up
five-pointed purple blooms, no larger than a nickel, spiked with
vibrant yellow stamens. Late season, the flowers give way to fruits
that could be mistaken for their agriculturally engineered cousins—
yellow grape tomatoes. But such a mistake could be fatal. The whole
plant is toxic, and livestock learn to avoid it, leading to its proliferation in overgrazed areas like this.

Still, even Guy B. Kyser, weed eradication expert at the University
of California, Davis, admits to its "sinister charm." It is admirably
stubborn. Armored in hairy spines, with long roots and a vanishing
need for water, it is nearly impossible to clear without pesticides.

It's a gift to find reverence for the small and the common.
Even in the driest summers, when the oversweet stink of Roundup
drifted on the wind, sprinklings of purple lined the churned margins of the dirt roadway and sprang from cracks in the sunbeaten
clay. Up close, I've noticed the translucent green hump of a tortoise beetle nudging its way down a leaf. Once, I was lucky enough
to see at sunset: sphinx moths riddling between the blossoms.

Couch's Spadefoot
(Scaphiopus Couchi)

Any dry place is colored by long desire. The annual dozen or so
inches of relief that fall in Midland seem to come all at once. Some
summer afternoon, I was nearly seduced into staying out watching
mammatus clouds, full as flesh pressed against fishnet, build into
towering anvils. But I thought better of it and retreated home to
listen to the pelting rain and hail.

In the days after, the ground let loose something resembling
conventional beauty.

The pencil stems of zephyr lilies poked up from the sand, opening into wax-white stars seemingly only hours after the rain. Pale bells of yucca flowers, dangling from their spires, dripped nectar that drew ants and the specialist moths evolved to pollinate them. Tiny explosions of violet Tahoka daisies, Spanish gold, and scarlet blanket flower erupted in patches that drew butterflies—papery sulfurs and tiny blues decorated like Persian miniatures. Ornate box turtles emerged from the brush to sip from short-lived puddles.

This abundance was almost enough to obscure waterlogged curls of carpeting scraps or the sopping mattress dumped in the bar ditch.

Along one stretch of my dirt road, the subterranean course of the gas pipeline is outlined by yellow pickets warning against digging, but all around them, another kind of delayed satisfaction emerged with the rain: spadefoot toads unburying themselves from months of torpor. Herpetologists believe that they're awakened not by moisture, but by the rumbling of thunder and precipitation striking the ground. In a few damp days (or even hours), they eat, find a tadpole-suitable source of water, mate, and—using toughened spurs on their rear feet—reburrow themselves.

Their pools must last ten days for spadefoot offspring to develop. A challenge, given that sporadic storms are often followed by air-fryer heat. Playa wetlands—natural depressions in which rain collects to form ephemeral ponds—are vanishing across the southern plains as they fill with sediment from upland erosion or are simply evaporated by an increasingly hotter, drier climate.

However, on the outskirts of housing developments, catchment basins provide an alternative. Channeling water away from neighborhood streets, ours filled just long enough to provide adequate mating ground. On its slopes, dozens of spadefoot males creaked enticements into an evening scented by chocolate daisies just blooming.

Dogweed, or Dahlberg Daisy (*Thymophylla Tenuiloba*)

In evaluating ecological aesthetics, we often reduce judgments to dichotomies—of ugliness and beauty, natural and unnatural—drawn from notions of purity. Canadian philosopher Alexis Shotwell identifies such distinctions as a legacy of colonialism. In describing contemporary appeals to the purity of the individual, she illustrates the absurdity of drawing the boundaries of our personal responsibility at the surface of our skin: "[Purity] means you don't have to feel bad about other people getting sick and dying because they're living downstream from a factory. They should've done something about that, they should've eaten more antioxidants."

Shit land, sitting as it does between developed and undeveloped territory, shows the absurdity of drawing such boundaries in the environment. In contaminated space, perceived ugliness, and therefore reduced ecological value, stem from impure status.

The tactics of environmentalism have conventionally been driven by two moves: to restore and to preserve. Land perceived as wounded is bandaged and nursed back to a healthy, ostensibly *natural* state, while land perceived as worthy of protection is demarcated as if placed beneath a glass cloche. But in increasingly unavoidable and obvious ways, these moves feel futile, even counterproductive, facing the challenges of our late-petrocene period.

Most patently, the boundless effects of climate change reduce any protective dome to gesture. This spring, the Rio Grande stopped flowing through swaths of its course in Big Bend National Park two hundred miles south of here. Meanwhile, to our west, New Mexico's Lincoln National Forest has seen a decade of destructive lightning strike fires fueled by extreme drought.

But a brief walk through a plot of shit land makes the interconnectedness of the natural world and human activity irrefutable. This place won't lie for us. And for that it should win our attention.

I knew from the start that there was an expiration date on my personal patch of shit land. Despite a couple of busts in the oil market, people haven't left. Advancing frack technology encourages continued drilling for oil in town. All this time, the edges have continued to be eroded by apartment buildings, a gas station, a Wendy's, more drilling, more subdivisions. The city paved the road and ripped up the windbreak where the jackrabbits hid. The foxes and owls are gone.

Soon, we'll be gone, too. My wife and I are moving to a neighborhood closer to "Old Midland," with canopies of street trees and a grassy public park.

It has often been remarked that West Texas smells like money. The dark, eggy scent of hydrogen sulfide and sulfur dioxide hangs on the air, even miles away from pumpjacks lifting sour crude. But the smell that I will always associate with this place is from a tiny plant with delicate yellow blooms and leathery foliage, too easy to step on while you have your eyes on something else. Crushed underfoot, it has an odor that comes back to me in dreams: sometimes like lemons and parsley, sometimes like a headache. Some people call it Dahlberg daisy, others dogweed. I could grow some in our new backyard, but I know that I won't. It doesn't belong there anyway.

B. M. OWENS

Lolita Floats Still in Miami

FROM *Hawai'i Pacific Review*

IMAGINE SWIMMING IN a pool. No, imagine living in that pool. Imagine that pool being all that exists in the world to you. The pool is your world and your world is 35 feet wide and 12 to 20 feet deep. You are 20 feet long and swim in constant circles as children bang on the see-through glass tank. High pitched whistles sound and you breach but you're not sure why. You're given food. That's why. You continue your circles, you're making something. The water laps around the sides. Your fins guide the water with incantations others don't understand—you don't really understand them either. You swim and swim and you're still here, swimming. A whirlpool forms at the center. This is it—You charge toward it, hoping the water sucks you in. That it'll tear holes into the bottom of the tank—into reality. That it'll pluck and sweep you into deep waters. That it'll bring you home out to the Pacific Ocean or, at least, drown you. But it doesn't. The water settles. Your body is stiff as you float beneath the Florida sun. Maybe if you're still enough the heat will melt your blubber and you can ooze out of here through the drains. The sun only blisters your skin but you don't seek shade because you already know there isn't any. This is all there is—this pool is your world.

Your name was Tokitae and you once dived almost 500 feet with your pod off Puget Sound but now you're here at Miami Seaquarium and they call you Lolita. You can smell the Atlantic Ocean. You can hear its sounds—laps of waves clapping against each other. Fish and their echoes of movement. You're in this pool but you can feel the sounds of an ocean you never knew. The gulls call out to you in a language that's familiar but forgotten. They try to tell you news of family on the other side of the continent. How your sisters and cousins have had multiple still born calves. How the dead bodies float at the surface, like the way you float in this pool. How their mothers carry them—push them with the tips of their noses. Gently prodding them to swim. The gulls tell you of the diminishing salmon supply. About the humans taking too much— the way they always take too much. The gulls tell you about the islands of trash they've seen. They try to tell you about the noise pollution. But their beaks can't make the sounds human sonar makes. And even if they could, you wouldn't understand. None of these sounds are familiar anymore.

You were taken from the Pacific Ocean at four years old and sold for $20,000 in 1970. You don't know what money is or what years are. There is no time keeping in your world—whales don't use money in wild or captive worlds. Except you're here to make money. You are a commodity. As long as you wave your flipper and breach and splash just enough. As long as the audience claps for you. As long as you don't attack the trainers, they'll make sure you're well fed.

The two white sided dolphins you share the tank with understand less than you do. But they care less because they have each other. You don't talk to them. If they swim close to you, you swim away. You haven't seen another orca in years. For all you know, you're the last orca in the world—you're at least the last orca in your world.

You shared the tank with Hugo, a male orca from Puget Sound, for the first ten years of captivity. He helped you remember your family. He brought the sounds of home here. He felt familiar. You'd fuck him constantly. What else is there to do in here? But you never had a calf of your own. You knew this was no place to raise a young whale. You knew they'd take your baby away from you. That your calf would grow up in another small tank. Your calf would be different from you—your calf would never know any ocean.

It was all too much and not enough for Hugo. Maybe you weren't enough for him. There's not enough of anything in here. He liked to ram his head into the tank. Maybe he thought if he hit the tank hard enough, he could shift its location on earth—tip the two of you into the Atlantic. Maybe he was creating a language—a pattern of bangs, calling out to kin. Maybe he thought he could expand the tank himself. Maybe he thought this would tell the humans that two orcas couldn't live here. One orca couldn't live here—this couldn't be your entire world. He kept ramming into the tank—into the hands of children banging on the viewing windows that have signs that say *please don't bang on the glass.* Once he almost tore off his rostrum—they had to surgically re-attach his upper jaw. Maybe he thought he could tear himself apart. Maybe he thought the pieces of him would find their way back to the ocean. His dorsal fin flopped over and he continued to bang against the tank as you coped by swimming in circles—unable to stop him. Unable to help him. Unable to do anything else.

Hugo died of a brain aneurism and they dumped his remains into the Miami-Dade landfill. Vultures picked at his flesh as the sun boiled away his blubber, sticking to fragments of plastic and cardboard and spoiled food. He lies in unrest in an ocean of trash and you don't know it, but that's where you'll end up too. Groups of activists are fighting in courts about your rights—they sign petitions. Protest. Post posters on billboards. Write essays. But even if they win, captive whales forget how to live in the ocean.

After two decades in captivity, Keiko, the whale actor in *Free Willy,* earned enough money from his captivity—he performed enough so they decided he could return home to Iceland. After years of therapy in a contained cove, where humans tried to teach him to catch his own fish again, they released him into the open ocean. He never found his family and he never socialized with other whales and he never remembered how to be a whale. In 2003, Keiko died of pneumonia. His body, at least, now rests on a quiet seabed inside the hearth of the ocean.

But you'll never pay off your debts. You're not a whale actress and your ticket sales are too low—if they were too high, Miami would never let you go. No, your entire world is this pool and there are no other fish or whales or oceans. So, you swim in circles, around the tank—casting circles of incantations. Spells to forget. Spiraling whirlpools surround you until you remember, again, this is your world. Stop—and float still.

*

Hawai`i Pacific Review published this essay on July 6, 2023, just a month before Tokitae (Lolita) would die of renal disease, pneumonia, and other causes in her tank at Miami Seaquarium on August 18, 2023. The Lummi (Lhaq'temish) Nation, who view the Southern Resident Orcas as relatives and had actively fought to bring Tokitae (Sk'aliCh'elh-tenaut) home, were not consulted before she was autopsied at the University of Georgia. A member of the Lummi Nation traveled to Georgia to culturally prepare Tokitae's ashes and bring her back home. On September 23, 2023, during a private Lummi water ceremony, they released Tokitae into the Salish Sea. Tokitae was the last of the Southern Resident Orcas still in captivity. As of December 2023, only seventy-four remain in the wild.

Nathan

FROM Longreads

I HAD BEEN volunteering at the ape house for four months before I was invited to meet Nathan. It was December and I'd just spent my first Christmas with the apes. Everyone but the director and I had left for the day. The night sky spilled over the glassceilinged, central atrium we called the greenhouse. Despite the snow outside, the greenhouse air was warm and ample. Moving toward the padlocked cage door, I felt light, as if I was about to float up into that dotted black expanse above me, rather than enter a room I'd cleaned feces and orange peels out of hours earlier.

I juggled my keys and the offering I'd brought with me—a tub of yogurt, a couple of bananas, Gatorade, and some blankets. With the two padlocks removed, I entered, sat, and arranged the gifts in an arc around me. Even though I was planted firmly on the glazed concrete floor, I swayed.

In the adjacent room, watching everything I did through the glass portion of a mechanical sliding door, was Nathan. Five years old to my twenty-one. He was stout, wide-shouldered, with thick muscled arms, but almost twiggy legs. Nathan was, simply put, a cool little dude. We studied one another as we waited for my supervisor to turn the key and remove the barrier between us. His eyes were as big and soft as three-quarter moons, always holding a question. Though, more often than not, that question was really a dare.

Sitting in the greenhouse, everything in the world was in alignment. It was right that Nathan would be the first ape I ever truly met. While the adult males still made crashing displays of warning at me, and the adult females mostly ignored me or found me to be a mildly useful, but mostly superfluous, component of the building, Nathan had always welcomed me warmly. I was a new playmate, willing to run back and forth along his enclosure in games of chase, over-eager to please.

A racking ka-chunk filled the greenhouse as the mechanical door separating us was activated. It rocked and then jerked to the side in its steel track. The doors had been created for use in prisons, originally, and they were always jamming on us. The third or fourth time we called a repairman out, he'd said we needed to take it easier on them—they hadn't been designed to open so often.

My breath stilled. I saw Nathan behind the door, then I saw the night sky. Nothing in between. I was on my back like some upturned turtle, my legs still crossed but now pointed at the stars. He was a heat-seeking missile. The impressions of both his feet were on my chest; the last breath I'd taken in a *Homo*-centric world evicted from my lungs. When I levered my body upright, I saw him waiting, peering at me from a foot or so away, head cocked. The air that rushed back into me was sweeter, lighter, than what had been there before.

"Hi," I said, grinning.

I guess I passed the test. He plopped himself in the bowl of my still-crossed legs, plucked the lid from the yogurt, and began to pour the thickness down his throat. He peeped contentedly and put his spare hand around the back of my neck. Where everything had been so fast that I couldn't take it in seconds before, now time was suspended. He smelled so clean, like construction paper and newly fallen leaves. We sat there, me running my fingers through the hair of his back, he slurping yogurt. Eventually, he pulled the blankets into a corner to construct his night nest. The director told me that meant it was bedtime. I told Nathan good night and we parted for the evening.

I'd initially applied to the ape house because I believed in their stated scientific mission: to communicate across the species boundary and illuminate the nonhuman and human mind. I had been one of those children at the zoo that try to make the right sound to get the animals to speak back. Now I was that kind of adult.

Not long after meeting and warming to the eight bonobos who were essentially my bosses, the science became much more personal. I was having trouble at college. My small rural campus felt like my cage. Though the school marketed itself as a home for outcasts looking for their place, I never felt entirely welcome there. I was shy and anxious to the level of needing therapy and medication (not receiving either), but I looked like a jock. When I did venture beyond my dorm room on weekends, I usually drank until I could approach and socialize with others (read: too much).

In the ape house, amongst the bonobos, I found the refuge my alma mater had promised to be. There, nothing rested on my ability to wrench words from my throat in front of my peers. In fact, my trend toward quiet was an asset while my athleticism was less intimidating than it was an invitation to play. For the first month, I tentatively hoped that the apes would have me. But after that month, had the humans offered me a room in one of the unused enclosures, I would have abandoned my degree and moved in with relief.

In my first weeks there, I asked my supervisor for tips on how to interact with them. "Just treat them like you treat me," she said. "Speak *to* them, not *about* them. Assume they're listening and can understand. They are and they can. These are people in nonhuman bodies and they know it." I could handle that. I was familiar with the fallacy that bodies accurately matched the selves they contained.

Likewise, it was comforting to be a part of a project that sought the person in the ape, to whatever degree it was present, rather than force the transformation. My research into the field showed that other ape language experiments had not been so accepting or accommodating. In the majority, the test subjects were taken from their mothers as infants and placed in human homes or labs. This was considered scientific. Rigorous. Rearing was the independent variable. To allow these subjects to remain with their birth families would be a confound.

While that December evening was my first time crossing the divide between *Pan* and *Homo,* Nathan had already been doing it for years. He was the third generation and fourth individual entrant into this ape language experiment. His upbringing was unusually casual for an experimental subject, and he spent nearly equal time with both his ape and human family. On a cultural spectrum from wild-caught bonobo (his grandmother and father), to human-reared language apes (his mother and brother), to human experimenter, Nathan sat exactly at the midpoint. He was the fulcrum upon which worlds balanced. The hope for him was that, under the direct tutelage of his mother, and with frequent but unstructured interactions with humans, he would show just how self-sustaining ape symbol use could be across generations. The avoidance of structure *was* the scientific methodology.

As poetic as I found it that Nathan was my point of first contact, he was simply the logical choice. He was small enough to handle— even if he was already stronger than me—and young enough that it was unlikely he'd attack should I misstep. Culturally, he was also optimally situated to understand my inexperience. He was an in- terpreter, an emissary. He was my bridge into the ape world.

I got no more training for being with Nathan than that first night. For every meeting thereafter, the only suggestion the direc- tor gave me was that I should always use the symbols—easily quanti- fied, discrete images. One per word. There were nouns, verbs, and even references to abstract concepts like time and feelings. The director thought maybe Nathan would help me learn them faster.

It seemed, at the time, that the only complication in Nathan's life was his big brother, Star, who was so perfect it was offensive. Star was irritatingly handsome, with a smile that smoothed over any and every slight. He spit on me daily but blew kisses to all the fe- male staff. Like many beautiful people, he was given credit for being smarter than he actually was and better behaved than he ever cared to be. Star's shadow was long and hard to escape. So, if Star showed an interest or proclivity for anything, Nathan either dismissed the activity outright or tried to do it harder/faster/better/stronger than anyone had ever done it before. The symbols were one of these things.

Nathan used the symbols like my father uses text messages, in- frequently, out of the blue, and with suspicious competency. I often caught Nathan in the corner of a room, his back to the door, symbol board in his lap. He'd be touching it, talking to himself. Thinking out loud, as it were. Other times, he'd saddle up before one of the touch-screen computer stations containing digital versions of the symbol board and rattle off a string of twenty or thirty symbols so fast the computer got bogged down in its processing and lagged in displaying them. I suspected he always meant exactly what he said, though I had no way of scientifically confirming this.

We ended up with a routine. I pretended that we were part of the experiment, doing important research, and he pretended not to understand what I was saying. A normal conversation between us using the symbols would look something like this:

Me: NATHAN YOU WANT FOOD, QUESTION?

Nathan somersaults into my lap, right over the symbol board.

Me (after extracting the board from under him): WANT FOOD, QUESTION?

Nathan pushes the board away. Hops up and runs away after biting me on my forearm. Playfully, but not without pain.

Me: I GET APPLES, QUESTION? GET CELERY, QUESTION? GET MILK, QUESTION?

Nathan approaches, holds my gaze from under his robust brow. I put the symbol board on the floor between us. He gestures, finger crooked, knuckle between his teeth. [Bite.]

Me: "Nathan, can you use the keyboard please?"
Nathan, hand snapping out: CHASE.

He springs away at full speed, a single fart helping propel him away down the hallway.

Me, following: "Okay, but no fair using rocket boosters."

I wasn't as diligent with the keyboard as I could have been, in part because I never had difficulty simply talking to him. In terms of receptive, rather than productive, competence, Nathan could handle it all. The rub was that he only listened when he felt like it. I often talked to him as I would any other person, except I was more honest and open. I started, genuinely, to consider Nathan one of my best friends.

He helped me work with the other apes, too. I would lay out maneuvers for shifting the apes between rooms and he would facilitate. He'd lead his family, including his grandmother, Worry, and his half-brother, Momo, through the door I'd indicated, then slip back through at the last moment, separating them in the new room while he and I got space to interact.

> Me: "Okay, here's the plan. Nathan, I want Worry and Momo to go to the greenhouse, but I want you to stay here so we can see each other and tickle and chase. Can you help me get them to move and you can stay here?"

Nathan peeps excitedly, and Worry and Momo echo him.

> Me: "They will have really good blueberries and lettuce and Gatorade in the greenhouse. We can have some surprises over here. Ready to help me move them? Okay, here we go."

Nathan sits by the door to the greenhouse, enthusiastic. He peeps to get the others interested. I operate the door and the others follow him into the transfer space between rooms. I start to close the door. At the last moment, Nathan slides through and sits alone in the room.

> Me: "Nathan, you did it. Great work, man."

Nathan runs to the mesh for a tummy tickle.

This went both ways, as the other apes used his skills, too. It was hell on data collection. I can't even count the number of times he ruined an experimental session because the non-language bonobos would drag him to the computer by the hand and wait while he performed their sets. He'd tap at the screen while they sat at the reward dispenser eating fruit chunk after fruit chunk produced for his correct answers.

One afternoon, after we had become full partners in crime, Nathan and I lounged in a pocket of space between the roof of the walk-in fridge and the kitchen ceiling. Sunlight floated lazily through the kitchen windows, warming the stainless steel of the countertops and cabinets, making the room toasty and our eyelids heavy. It was late spring, months since we'd first met, though it felt longer. Something about being with the apes made time less distinct.

When it was me and Nathan together, I could forget I was an employee and Nathan essentially my work. Our relationship had grown through months of one-on-one encounters. With each visit, we gained new privileges until there was hardly an inch of the building not available to us, so long as it wasn't occupied. It could just be me and my friend. He, a boy, and me, his cool but slightly irresponsible guardian. Gone were my problems at college. Gone were the impenetrable complexities of human relationships. My anxiety around humans was inversely proportional to my comfort in the cage with the bonobos. Apes made so much more sense to me, Nathan most of all. It eventually got to the point where I stopped going to school, seven credits short of a degree, to work with the apes full-time.

Between us in our nest atop the fridge was a pile of empty Diet Coke bottles, Go-Gurt tubes, and half a bag of plump, red grapes. There were plenty of vegetables in the fridge under us, but they held little appeal. When Nathan and I went to the kitchen, we were raiders. We descended like locusts and went straight for the good stuff.

The kitchen was our favorite place to go. It held not just food, but choice. There, Nathan could eat whatever he wanted, not what was brought to him by a caretaker. However, the kitchen was, ultimately, a human place, and as a result, I wasn't able to fully relax. There were all these reminders of how human spaces were not made to accommodate us. Blenders with stainless steel blades, kitchen knives, toxic cleaning agents, gas stove burners. Dangers everywhere.

Nathan dropped the last Diet Coke bottle between us and burped. I retrieved a paper board with the symbols on it. "Nathan," I asked, pointing to symbols to accompany my words.

WANT MORE COKE? WANT APPLE?

He pushed the board away, then pulled me in for a hug and tickle. If anything, Nathan taught me how impossible the science of ape language was to perform. His whole body was an instrument of expression. He manipulated the space between us like prose, varying the pressure of his teeth on my skin to change the tone of a message, his every touch held its own grammar as questions and statements. Nathan didn't perform language in a way that would be easy to parse and study, he embodied it. He performed it in the way of a dancer. He lived it.

Nathan preferred gestures. Words filled him up and he had to expel their captive energy through his limbs in a way the symbols couldn't facilitate. Crooked index finger between his teeth: [Bite me]. Point at keys hanging from my belt loop on a carabiner:

[Keys/Open]. Crooked middle finger twisting at a door: [Open/ Unlock]. Hand raised to his neck, motioning as if to let steam out of an Oxford shirt: [Collar].

If he gestured for a collar, I'd ask, "You want to go outside?" Or "You want to go to the kitchen?" He would vocalize in response, then sit with his chin raised to expedite the process. I didn't really like the collar, but whenever I could, I looped the thick nylon strap around his neck and locked the full-sized padlock that secured it. The heavy pendant hung between the ends of his collarbones. He inspected it with his fingers, adjusting to its heft. The thing was incongruous with this person, this child.

He asked for it every day I saw him. Often repeatedly. Switching between that gesture and the one asking for my keys. He wanted, more than just about anything else, to traverse the boundaries between ape and human space. For every step I took into his world, he was equally desperate—more so, even—to take one into mine. Every time I successfully begged, cajoled, and (sometimes) argued with humans for the opportunity to enter his world, he would greet me by asking for me to take him back to where I'd come from. Get me out of this place, he seemed to say.

So, I traded my discomfort with the collar for the chance to make him happy. He traded the cage he lived in for the one he wore around his neck. The easiest days were the ones when I didn't have to say "no" to him. When he asked for keys or a collar and I could say "Of course" and we would go gorge ourselves and loiter on top of the walk-in fridge.

I lived for those days of forgetting. The times when we found the right balance between the demands of our worlds and our own desires, but I was lucky if there was enough staff to accommodate us having half the building once every few weeks or so.

Though Nathan had been raised to be both bonobo and human, his was a secondary type of personhood. Not like that of a human child. He could enter the kitchen, but only on a leash. He was taught, but could not go to school. He had the language to ask to go outside, but he could never venture beyond the walls of the facility. I kept trying to find ways to make up for that disconnect, but, as a frustratingly junior member of staff, I couldn't.

One day we lazed on top of the fridge until Nathan stirred and descended. My thoughts came slow in the sun-warmed room. I thought he wanted a different kind of snack until he moved toward the sink. His head disappeared as he ducked under with his leash dragging behind him.

"Nathan, c'mon, man," I said. "Nothing good down there."

I scrambled down, imagining a montage of him ingesting jugs of cleaning solvents or blinding his eyes with toxic sprays. I approached but before I could reach him, Nathan hung from the sink lip, reared back, and kicked the garbage disposal with all his considerable muscle. He planted several rocking blows to it before I got him turned around.

The spell he cast that made me forget the human world dissipated with the thuds of his feet against metal. I was a human and, worse, an employee. He was an ape then. It hurt to be reminded of that.

I didn't want to get in trouble. I couldn't afford to replace the garbage disposal. Worse, I couldn't afford to have my time with Nathan revoked or reduced to less than it already was. But even more than that, I wanted to prove that we had something. That our connection was real and tangible. I knew he was special, but I wanted *us* to be special too.

I pulled him away from the sink, my ears hot. He'd never been so blatantly destructive around me before.

"What's wrong with you?" I used the voice I give to my dogs when they misbehave. "No!"

Nathan didn't meet my eyes. He squirmed away only to plant another rocking blow on the disposal. I pulled him back into position with firm hands on his shoulders.

"No, Nathan. No! That's bad." I was near to shouting.

Nathan's eyes were hard at the corners. He tested my hold once more, paused, then opened his mouth and screamed. He wailed long strings of ear-splitting EEEs. The whole ape house heard him. They barked, sharp, in response. He screamed so hard and so much that within minutes all his skin had broken out in half-dollar-sized hives. I unhanded him and he left my side to go sulk in a corner, screaming all the while.

The director, who'd heard the commotion, joined us after a few minutes. Nathan sprang into her arms and hugged her close, looking at me the whole time. Using his proximity to her and distance from me to express his displeasure. She soothed him and I explained the situation. Before she returned him to his ape family with a dose of liquid grape children's Benadryl in him, I apologized. I gave him some M&M's and a special juice box and, after a pause, he offered his back for a tickle. He would accept my offering, but he wanted me to be sorry for longer.

"Disagreements," the director said after returning him, "are part of having language."

The hives were no surprise to her. Nathan often got so worked up that his body rebelled. As if his emotions, same as his words, were too strong for their little container and pushed against his skin to escape.

They were the main reason why we didn't notice when he got actually sick.

The study of ape language is a field of broken promises. Its history is littered with the allegedly well-meaning intentions of seemingly caring people and the tragic, too early passings of their charges. Their failures are made all the more devastating in that, despite what they call the apes—subjects, participants, entrants—they are the failures of parents toward their children.

Ape language research, at its heart, seeks to investigate the age-old question of nature versus nurture. Since raising human infants in a context removed from all human influence is ethically impossible, they performed the inverse: raising apes in entirely human environments. That has historically not met the same ethical barriers, despite infants being involved in each case.

Take Gua, adopted by a pair of psychologists, the Kelloggs, who had recently given birth to their first child, Donald. The Kelloggs stressed that for any co-rearing experiment to work, the ape must be treated as human in all regards, to avoid bias. As such, Gua lived in the Kelloggs' house, ate at the table, and generally did everything with Donald. They were inseparable, like twins, and they developed at almost identical rates in everything but their speech.

Winthrop Kellogg's original hypothesis, that Gua would develop aspects of human behavior, proved true. What he did not anticipate, however, was that this cultural blending would be a two-way exchange; the spectrum between *Pan* and *Homo* traversed in both directions. While Gua grew more human, Donald also took on some of Gua's apeness, such as extensive biting. The two children met in the middle, the primary contributor to the end of the experiment.

In each ape language study, there is one overriding and unspoken promise—we will give you a new family *if* you become sufficiently like us (but not if our children become like you). Unlike Gua and others, Nathan kept his ape family. Still, the promise of his life was the same, if the terms slightly altered. It all boiled down to this: We will make you one of us.

No study has yet been able to make good on that promise. No matter what the shape, be it a collar, a mesh enclosure, or a house, there is always a cage around the apes involved. They are never truly welcomed into human society. The humans, meanwhile, get to go home at the end of the day.

Of the approximately hundred years of other ape language studies, hardly any of the apes had Nathan's freedom. These apes were almost all taken from their mothers as infants, some as young as two days old, and placed in human homes or labs. Nearly all lived short, tragic lives compared to the potential sixty to seventy years available to them naturally. As if a stark contrast between the mental and physical self invariably tempts tragedy. Kellogg's Gua was returned to a research center after nine months in their home (pneumonia, three); Ladygina-Kohts' Joni ate paint from the walls of her home (lead poisoning, three); Hayes' Viki fell sick during the study period (viral meningitis, seven); Temerlin's Lucy was released back into the wild after living in a human home in Oklahoma for years (suspected poaching, twenty-three); and Nim Chimpsky was "retired" to a research lab, which sold him into biomedical research, from which he was rescued by an animal sanctuary to live as their sole chimpanzee (heart attack, twenty-six). So many either didn't survive their studies or barely did. The handful of language apes, like Washoe, Koko, and Kanzi who have lived into and beyond their third decade are rare exceptions.

Several months after the garbage disposal disagreement, factors outside our control interrupted our time together. In the human half of the building, new leadership took over, stiffening the rules about contact with apes. Months passed. Then, just as we were about to renew our one-on-ones, Nathan got sick. I saw him daily during this time but it was always through the mesh of the cage. I pushed so far through it to touch and tickle him that it hurt the web of skin between my fingers.

It came out of nowhere. One day his face just swelled up. His eyes shrank to crescents between his puffy brow and cheeks. No one had any answers for it, not even the vet. Every morning I came in, Nathan looked like he had been in a boxing match the night before. We gave him Benadryl and Claritin over and over. It made him groggy, but it didn't seem to help his swelling. Nothing seemed to help. We eliminated potential allergens. Changed cleaning solutions, avoided wheat gluten, and banned food with certain dyes. All to no effect.

As the sickness swept through him, he maintained a front of normalcy. When he chose to talk using symbols, it still came out in torrents. When he wanted to chase, he ran as fast as he could, even if the run was more of a tumble and the game didn't last as long as usual. The vet, whose practice focused primarily on Iowa farm animals, visited often. She did her best, but Nathan was a boy, not a horse.

It didn't go away. I asked that he see a different doctor, a human one. But in this, he was not human enough. There were ape-specific risks of a more thorough workup and, it was assumed, they outweighed the benefits given his symptoms. A full workup would require sedation and transportation and more. Nathan's father had died two years earlier from complications with anesthesia for an elective procedure and his loss was still fresh in everyone's heart.

Over the course of half a year, Nathan's swelling receded as mysteriously as it had arisen. By the time spring rolled around, he was almost normal, though his hair was a little wirier and his arms had lost some of their beef. His eyes also drooped at the outside, making him look eternally tired. But he was nearly his old self, if more subdued.

By May, with the fields outside bursting with purple, orange, and yellow wildflowers, I finally got the supervisor's approval to go in again with Nathan. I'd been requesting it for months. Just after I got the green light, however, Nathan stopped eating and our reunion was put on hold. It didn't matter the meal, he took a couple of bites and set it aside. Then his breathing became labored. He wheezed and coughed so loud I could hear it throughout the building. His energy gone, he spent most of his time napping. I knew I had to see him, so I did.

Nathan was dozing when I entered his room for the first time in over a year. It was late morning. I didn't ask permission, I simply told the other caretakers that I would need that half of the building.

"Hi, Nathan," I said as I entered. He was lying on a pile of blankets. He didn't move at the sound of the door, but as I spoke, he lifted himself and approached. The slump of his shoulders told me just how uncomfortable he was. His swagger was gone. I didn't think anything could take the strut out of his walk. Now, he was deflated. He hadn't eaten more than a couple of bites in days.

And yet, he didn't miss a beat. He hugged me about the legs, slapped my thighs, and sprung away awkwardly. Just like we normally greeted one another, only in slow motion. Now his sprint was more of a lope. I shuffled so I didn't overtake him. We did one round of this before he led me back to his bed, laid down, and asked me to tickle him. As my fingers probed his ribs, he grunted a laugh that became wheezing and quickly turned into a racking cough. It passed, and he looked at me with his mouth hanging slightly agape as if all the strength required to close it was concentrated at the corners of his wincing eyes. I began to tickle him again, this time softly, but he brushed my hands away.

I shouldn't have let so much time go by, I thought as we sat there, my back to one wall and Nathan inert across my thighs. It used to take hours before he'd slow down enough that we could relax like this. Today it took barely a minute.

My fingers tentatively massaged him. They met bone much easier than before. The curving mounds of his muscles were reduced, his skin slack. During the worst of his sickness, when the swelling and itching were at their highest, he'd pulled most of his arm hair out. The baldness highlighted his new angularity. I ran my fingers over his bare forehead. His sideburns were plucked clean and what hair was left was brittle stubble, bending and snapping like sun-bleached grass.

Someone brought a scale to get Nathan's weight for the vet. He didn't want to move and threatened to bite me when I suggested it. I waited a minute for him to doze off again, then picked him up and carried him to the scale. He'd lost over 20 pounds in under three months.

We spent the remainder of the day resting. With me running my hand over his skin, and him in a near-constant adjustment of his position. Intermittently, I'd leave to get him a Popsicle or some juice. I took one of his bare feet in my hand and nibbled on his toes. He huffed one laugh as if to humor me, but nothing more. I brought him M&M's, but these were too hard for him to eat and he set them aside.

That night, I entered his cage with fresh blankets and a bowl of yogurt, an echo of our first meeting. He tried a bit of yogurt, then put the bowl down next to his bed. I'd been asked to get a blood oxygen reading for the vet with a clip that went on the end of Nathan's finger. I moved to his side while he slipped in and out of an uneasy sleep and took his hand in mine. A coworker threaded the sensor through the mesh. Before I could clamp the device on his index finger, he woke, lunging and snapping at me. He didn't get me, but the anger in his movements stung as much as a bite would have.

I felt like I was betraying him, putting human obligations above his very clear refusal. He let me hold his hand again. This time I just held it. When he seemed to be fully asleep, I tried again. Once it got a reading, I unclamped it quickly, whispered good night, and slunk out of his room.

Eight hours later, he was carried out of the building on a blanket, finally breaking free of its walls, to get a full medical workup. During the night he had briefly gone into respiratory arrest. The risks of getting him checked out were now outweighed by the seriousness of his condition. They carried him by me, sleeping, but with his hands curled and ready, thumbs against the ends of his drawn index and middle fingers. I saw the potential in them. They were poised as if ready to ask for his *Collar* or my *Keys* at the very instant he woke.

In the years since I have often wondered what we accomplished in the ape house. What exactly was it that I was a part of? Did those in charge really believe all that they were saying? I thought we were doing it better, in knowing no one ever needed to tear infant apes from their mothers to learn about the limits of language. The other ape language studies had got the question wrong, I thought. They all asked whether an ape could talk if we made them sufficiently in our image. I thought we were asking if we could understand each other as equals. The true test not being in the apes' ability to speak but in our capacity to listen.

I thought we were different. Better. But, we were not, our bonobos no more equal than the charges of any other study. Our cages and facilities were simply nicer; our methods softer.

So much of my understanding of language, and its limits, came from Nathan. His silences especially. Language is messy and incomplete and variable and profound and decidedly unscientific. There is no single, controllable, independent variable. After all, there are so many things that are beyond the ability of words to express. So much meaning outside that which is merely spoken.

For example: There was no symbol for CANCER on the symbol keyboard. No one had ever needed to say LYMPHOMA. The lexicon was limited, but HURT was there, and I had never once seen Nathan use it. It wasn't that he didn't understand, it was that he would never admit such a thing. He had too big a chip on his shoulder.

For example: The way my coworker's voice caught on the phone, starting several utterances until "He didn't make it" could escape, and I had already known what he had to tell me. And the way I made the same stutter-stop code of not-quite-shock and not-yet-loss before managing "I'm on my way" in response, and he had already known that as well.

For example: How the people I passed as I walked through the ape house, hood over my head, made soft, unintelligible noises at me. Emitting contributions to the pall over the building. I kept moving, unsure of whether a response was expected. Unable to make one if it was. I just continued walking toward the van that had taken him to the hospital and back, parked at the other end of the facility.

For example: In the van—the gray—the interior gray—sky gray—world gray—the cold of his hand—he—splayed—the coolness of his forehead—kissing the stubble of his forehead—kissing and muttering—same three syllables—waiting for warmth to return.

For example: The stillness of the building as he was carried in and laid before the glass of the greenhouse where his family waited, pressed against the window, shoulders one against another, crowding together. The silence as deep and absolute as the understanding in his mother's eyes.

ELIZABETH KOLBERT

Talk to Me

FROM *The New Yorker*

> Ah, the world! Oh, the world!
> —*Moby-Dick*

DAVID GRUBER BEGAN his almost impossibly varied career studying bluestriped grunt fish off the coast of Belize. He was an undergraduate, and his job was to track the fish at night. He navigated by the stars and slept in a tent on the beach. "It was a dream," he recalled recently. "I didn't know what I was doing, but I was performing what I thought a marine biologist would do."

Gruber went on to work in Guyana, mapping forest plots, and in Florida, calculating how much water it would take to restore the Everglades. He wrote a Ph.D. thesis on carbon cycling in the oceans and became a professor of biology at the City University of New York. Along the way, he got interested in green fluorescent proteins, which are naturally synthesized by jellyfish but, with a little gene editing, can be produced by almost any living thing, including humans.

While working in the Solomon Islands, northeast of Australia, Gruber discovered dozens of species of fluorescent fish, including a fluorescent shark, which opened up new questions. What would a fluorescent shark look like to another fluorescent shark? Gruber enlisted researchers in optics to help him construct a special "shark's eye" camera. (Sharks see only in blue and green; fluorescence, it turns out, shows up to them as greater contrast.) Meanwhile, he was also studying creatures known as comb jellies at the Mystic Aquarium, in Connecticut, trying to determine how, exactly, they manufacture the molecules that make them glow.

This led him to wonder about the way that jellyfish experience the world. Gruber enlisted another set of collaborators to develop robots that could handle jellyfish with jellyfish-like delicacy.

"I wanted to know: Is there a way where robots and people can be brought together that builds empathy?" he told me.

In 2017, Gruber received a fellowship to spend a year at the Radcliffe Institute for Advanced Study, in Cambridge, Massachusetts. While there, he came across a book by a free diver who had taken a plunge with some sperm whales. This piqued Gruber's curiosity, so he started reading up on the animals.

The world's largest predators, sperm whales spend most of their lives hunting. To find their prey—generally squid—in the darkness of the depths, they rely on echolocation. By means of a specialized organ in their heads, they generate streams of clicks that bounce off any solid (or semi-solid) object. Sperm whales also produce quick bursts of clicks, known as codas, which they exchange with one another. The exchanges seem to have the structure of conversation.

One day, Gruber was sitting in his office at the Radcliffe Institute, listening to a tape of sperm whales chatting, when another fellow at the institute, Shafi Goldwasser, happened by. Goldwasser, a Turing Award–winning computer scientist, was intrigued. At the time, she was organizing a seminar on machine learning, which was advancing in ways that would eventually lead to ChatGPT. Perhaps, Goldwasser mused, machine learning could be used to discover the meaning of the whales' exchanges.

"It was not exactly a joke, but almost like a pipe dream," Goldwasser recollected. "But David really got into it."

Gruber and Goldwasser took the idea of decoding the codas to a third Radcliffe fellow, Michael Bronstein. Bronstein, also a computer scientist, is now the DeepMind Professor of A.I. at Oxford.

"This sounded like probably the most crazy project that I had ever heard about," Bronstein told me. "But David has this kind of power, this ability to convince and drag people along. I thought that it would be nice to try."

Gruber kept pushing the idea. Among the experts who found it loopy and, at the same time, irresistible were Robert Wood, a roboticist at Harvard, and Daniela Rus, who runs M.I.T.'s Computer Science and Artificial Intelligence Laboratory. Thus was born the Cetacean Translation Initiative—Project CETI for short. (The acronym is pronounced "setty," and purposefully recalls SETI, the Search for Extraterrestrial Intelligence.) CETI represents the most ambitious, the most technologically sophisticated, and the most well-funded effort ever made to communicate with another species.

"I think it's something that people get really excited about: Can we go from science fiction to science?" Rus told me. "I mean, can we talk to whales?"

Sperm whales are nomads. It is estimated that, in the course of a year, an individual whale swims at least twenty thousand miles. But scattered around the tropics, for reasons that are probably squid-related, there are a few places the whales tend to favor. One of these is a stretch of water off Dominica, a volcanic island in the Lesser Antilles.

CETI has its unofficial headquarters in a rental house above Roseau, the island's capital. The group's plan is to turn Dominica's west coast into a giant whale-recording studio. This involves installing a network of underwater microphones to capture the codas of passing whales. It also involves planting recording devices on the whales themselves—cetacean bugs, as it were. The data thus collected can then be used to "train" machine-learning algorithms.

In July, I went down to Dominica to watch the CETI team go sperm-whale bugging. My first morning on the island, I met up with Gruber just outside Roseau, on a dive-shop dock. Gruber, who is fifty, is a slight man with dark curly hair and a cheerfully anxious manner. He was carrying a waterproof case and wearing a CETI T-shirt. Soon, several more members of the team showed up, also carrying waterproof cases and wearing CETI T-shirts. We climbed aboard an oversized Zodiac called *CETI 2* and set off.

The night before, a tropical storm had raked the region with gusty winds and heavy rain, and Dominica's volcanic peaks were still wreathed in clouds. The sea was a series of white-fringed swells. *CETI 2* sped along, thumping up and down, up and down. Occasionally, flying fish zipped by; these remained aloft for such a long time that I was convinced for a while they were birds.

About two miles offshore, the captain, Kevin George, killed the engines. A graduate student named Yaly Mevorach put on a set of headphones and lowered an underwater mike—a hydrophone—into the waves. She listened for a bit and then, smiling, handed the headphones to me.

The most famous whale calls are the long, melancholy "songs" issued by humpbacks. Sperm-whale codas are neither mournful nor musical. Some people compare them to the sound of bacon frying, others to popcorn popping. That morning, as I listened through the headphones, I thought of horses clomping over cobbled streets. Then I changed my mind. The clatter was more mechanical, as if somewhere deep beneath the waves someone was pecking out a memo on a manual typewriter.

Mevorach unplugged the headphones from the mike, then plugged them into a contraption that looked like a car speaker riding a broom handle. The contraption, which I later learned had been jury-rigged out of, among other elements, a metal salad bowl, was designed to locate clicking whales. After twisting it around in the water for a while, Mevorach decided that the clicks were coming from the southwest. We thumped in that direction, and soon George called out, "Blow!"

A few hundred yards in front of us was a gray ridge that looked like a misshapen log. (When whales are resting at the surface, only a fraction of their enormous bulk is visible.) The whale blew again, and a geyser-like spray erupted from the ridge's left side.

As we were closing in, the whale blew yet again; then it raised its elegantly curved flukes into the air and dove. It was unlikely to resurface, I was told, for nearly an hour.

We thumped off in search of its kin. The farther south we traveled, the higher the swells. At one point, I felt my stomach lurch and went to the side of the boat to heave.

"I like to just throw up and get back to work," Mevorach told me.

Trying to attach a recording device to a sperm whale is a bit like trying to joust while racing on a Jet Ski. The exercise entails using a thirty-foot pole to stick the device onto the animal's back, which in turn entails getting within thirty feet of a creature the size of a school bus. That day, several more whales were spotted. But, for all of our thumping around, *CETI 2* never got close enough to one to unhitch the tagging pole.

The next day, the sea was calmer. Once again, we spotted whales, and several times the boat's designated pole-handler, Odel Harve, attempted to tag one. All his efforts went for naught. Either the whale dove at the last minute or the recording device slipped off the whale's back and had to be fished out of the water. (The device, which was about a foot long and shaped like a surfboard, was supposed to adhere via suction cups.) With each new sighting, the mood on *CETI 2* lifted; with each new failure, it sank.

On my third day in Dominica, I joined a slightly different subset of the team on a different boat to try out a new approach. Instead of a long pole, this boat—a forty-foot catamaran called *CETI 1*—was carrying an experimental drone. The drone had been specially designed at Harvard and was fitted out with a video camera and a plastic claw.

Because sperm whales are always on the move, there's no guarantee of finding any; weeks can go by without a single sighting off Dominica. Once again, though, we got lucky, and a whale was soon spotted. Stefano Pagani, an undergraduate who had been brought along for his piloting skills, pulled on what looked like a VR headset, which was linked to the drone's video camera. In this way, he could look down at the whale from the drone's perspective and, it was hoped, plant a recording device, which had been loaded into the claw, on the whale's back.

The drone took off and zipped toward the whale. It hovered for a few seconds, then dropped vertiginously. For the suction cups to adhere, the drone had to strike the whale at just the right angle, with just the right amount of force. Post impact, Pagani piloted the craft back to the boat with trembling hands. "The nerves get to you," he said.

"No pressure," Gruber joked. "It's not like there's a *New Yorker* reporter watching or anything." Someone asked for a round of applause. A cheer went up from the boat. The whale, for its part, seemed oblivious. It lolled around with the recording device, which was painted bright orange, stuck to its dark-gray skin. Then it dove.

Sperm whales are among the world's deepest divers. They routinely descend two thousand feet and sometimes more than a mile. (The deepest a human has ever gone with scuba gear is just shy of eleven hundred feet.) If the device stayed on, it would record any sounds the whale made on its travels. It would also log the whale's route, its heartbeat, and its orientation in the water. The suction was supposed to last around eight hours; after that—assuming all went according to plan—the device would come loose, bob to the surface, and transmit a radio signal that would allow it to be retrieved.

I said it was too bad we couldn't yet understand what the whales were saying, because perhaps this one, before she dove, had clicked out where she was headed.

"Come back in two years," Gruber said.

Every sperm whale's tail is unique. On some, the flukes are divided by a deep notch. On others, they meet almost in a straight line. Some flukes end in points; some are more rounded. Many are missing distinctive chunks, owing, presumably, to orca attacks. To ID a whale in the field, researchers usually rely on a photographic database called Flukebook. One of the very few scientists who can do it simply by sight is CETI's lead field biologist, Shane Gero.

Gero, who is forty-three, is tall and broad, with an eager smile and a pronounced Canadian accent. A scientist-in-residence at Ottawa's Carleton University, he has been studying the whales off Dominica since 2005. By now, he knows them so well that he can relate their triumphs and travails, as well as who gave birth to whom and when. A decade ago, as Gero started having children of his own, he began referring to his "human family" and his "whale family." (His human family lives in Ontario.) Another marine biologist once described Gero as sounding "like Captain Ahab after twenty years of psychotherapy."

When Gruber approached Gero about joining Project CETI, he was, initially, suspicious. "I get a lot of e-mails like 'Hey, I think whales have crystals in their heads,' and 'Maybe we can use them to cure malaria,'" Gero told me. "The first e-mail David sent me was like, 'Hi, I think we could find some funding to translate whale.' And I was like, 'Oh, boy.'"

A few months later, the two men met in person, in Washington, D.C., and hit it off. Two years after that, Gruber did find some funding. CETI received thirty-three million dollars from the Audacious Project, a philanthropic collaborative whose backers include Richard Branson and Ray Dalio. (The grant, which was divided into five annual payments, will run out in 2025.)

The whole time I was in Dominica, Gero was there as well, supervising graduate students and helping with the tagging effort. From him, I learned that the first whale I had seen was named Rita and that the whales that had subsequently been spotted included Raucous, Roger, and Rita's daughter, Rema. All belonged to a group called Unit R, which Gero characterized as "tightly and actively social." Apparently, Unit R is also warmhearted. Several years ago, when a group called Unit S got whittled down to just two members—Sally and TBB—the Rs adopted them.

Sperm whales have the biggest brains on the planet—six times the size of humans'. Their social lives are rich, complicated, and, some would say, ideal. The adult members of a unit, which may consist of anywhere from a few to a few dozen individuals, are all female. Male offspring are permitted to travel with the group until they're around fifteen years old; then, as Gero put it, they are "socially ostracized." Some continue to hang around their mothers and sisters, clicking away for months unanswered. Eventually, though, they get the message. Fully grown males are solitary creatures. They approach a band of females—presumably not their immediate relatives—only in order to mate. To signal their arrival, they issue deep, booming sounds known as clangs. No one knows exactly what makes a courting sperm whale attractive to a potential mate; Gero told me that he had seen some clanging males greeted with great commotion and others with the cetacean equivalent of a shrug.

Female sperm whales, meanwhile, are exceptionally close. The adults in a unit not only travel and hunt together; they also appear to confer on major decisions. If there's a new mother in the group, the other members mind the calf while she dives for food. In some units, though not in Unit R, sperm whales even suckle one another's young. When a family is threatened, the adults cluster together to protect their offspring, and when things are calm the calves fool around.

"It's like my kids and their cousins," Gero said.

The day after I watched the successful drone flight, I went out with Gero to try to recover the recording device. More than twenty-four hours had passed, and it still hadn't been located. Gero decided to drive out along a peninsula called Scotts Head, at the southwestern tip of Dominica, where he thought he might be able to pick up the radio signal. As we wound around on the island's treacherously narrow roads, he described to me an idea he had for a children's book that, read in one direction, would recount a story about a human family that lives on a boat and looks down at the water and, read from the other direction, would be about a whale family that lives deep beneath the boat and looks up at the waves.

"For me, the most rewarding part about spending a lot of time in the culture of whales is finding these fundamental similarities, these fundamental patterns," he said. "And, you know, sure, they won't have a word for 'tree.' And there's some part of the sperm-whale experience that our primate brain just won't understand. But those things that we share must be fundamentally important to why we're here."

After a while, we reached, quite literally, the end of the road. Beyond that was a hill that had to be climbed on foot. Gero was carrying a portable antenna, which he unfolded when we got to the top. If the recording unit had surfaced anywhere within twenty miles, Gero calculated, we should be able to detect the signal. It occurred to me that we were now trying to listen for a listening device. Gero held the antenna aloft and put his ear to some kind of receiver. He didn't hear anything, so, after admiring the view for a bit, we headed back down. Gero was hopeful that the device would eventually be recovered. But, as far as I know, it is still out there somewhere, adrift in the Caribbean.

The first scientific, or semi-scientific, study of sperm whales was a pamphlet published in 1835 by a Scottish ship doctor named Thomas Beale. Called *The Natural History of the Sperm Whale*, it proved so popular that Beale expanded the pamphlet into a book, which was issued under the same title four years later.

At the time, sperm-whale hunting was a major industry, both in Britain and in the United States. The animals were particularly prized for their spermaceti, the waxy oil that fills their gigantic heads. Spermaceti is an excellent lubricant, and, burned in a lamp, produces a clean, bright light; in Beale's day, it could sell for five times as much as ordinary whale oil. (It is the resemblance between semen and spermaceti that accounts for the species' embarrassing name.)

Beale believed sperm whales to be silent. "It is well known among the most experienced whalers that they never produce any nasal or vocal sounds whatever, except a trifling hissing at the time of the expiration of the spout," he wrote. The whales, he said, were also gentle—"a most timid and inoffensive animal." Melville relied heavily on Beale in composing *Moby-Dick*. (His personal copy of *The Natural History of the Sperm Whale* is now housed in Harvard's Houghton Library.) He attributed to sperm whales a "pyramidical silence."

"The whale has no voice," Melville wrote. "But then again," he went on, "what has the whale to say? Seldom have I known any profound being that had anything to say to this world, unless forced to stammer out something by way of getting a living."

The silence of the sperm whales went unchallenged until 1957. That year, two researchers from the Woods Hole Oceanographic Institution picked up sounds from a group they'd encountered off the coast of North Carolina. They detected strings of "sharp clicks," and speculated that these were made for the purpose of echolocation. Twenty years elapsed before one of the researchers, along with a different colleague from Woods Hole, determined that some sperm-whale clicks were issued in distinctive, often repeated patterns, which the pair dubbed "codas." Codas seemed to be exchanged between whales and so, they reasoned, must serve some communicative function.

Since then, cetologists have spent thousands of hours listening to codas, trying to figure out what that function might be. Gero, who wrote his Ph.D. thesis on vocal communication between sperm whales, told me that one of the "universal truths" about codas is their timing. There are always four seconds between the start of one coda and the beginning of the next. Roughly two of those seconds are given over to clicks; the rest is silence. Only after the pause, which may or may not be analogous to the pause a human speaker would put between words, does the clicking resume.

Codas are clearly learned or, to use the term of art, socially transmitted. Whales in the eastern Pacific exchange one set of codas, those in the eastern Caribbean another, and those in the South Atlantic yet another. Baby sperm whales pick up the codas exchanged by their relatives, and before they can click them out proficiently they "babble."

The whales around Dominica have a repertoire of around twenty-five codas. These codas differ from one another in the number of their clicks and also in their rhythms. The coda known as three regular, or 3R, for example, consists of three clicks issued at equal intervals. The coda 7R consists of seven evenly spaced clicks. In seven increasing, or 7I, by contrast, the interval between the clicks grows longer; it's about five-hundredths of a second between the first two clicks, and between the last two it's twice that long. In four decreasing, or 4D, there's a fifth of a second between the first two clicks and only a tenth of a second between the last two. Then, there are syncopated codas. The coda most frequently issued by members of Unit R, which has been dubbed 1 + 1 + 3, has a cha-cha-esque rhythm and might be rendered in English as click . . . click . . . click-click-click.

If codas are in any way comparable to words, a repertoire of twenty-five represents a pretty limited vocabulary. But, just as no one can yet say what, if anything, codas mean to sperm whales, no one can say exactly what features are significant to them. It may be that there are nuances in, say, pacing or pitch that have so far escaped human detection. Already, CETI team members have identified a new kind of signal—a single click—that may serve as some kind of punctuation mark.

When whales are resting near the surface, their exchanges can last an hour or more. Even by human standards, sperm-whale chatter is insistent and repetitive. "They're talking on top of each other all the time," Gero told me.

A snatch of dialogue recorded between two members of Unit R runs as follows. (Both Roger and Rita are adult females.)

    ROGER: 1 + 1 + 3
    RITA: 1 + 1 + 3, 1 + 1 + 3
    ROGER: 9I
    RITA: 1 + 1 + 3
    ROGER: 10I
    RITA: 1 + 1 + 3, 1 + 1 + 3
    ROGER: 11I
    RITA: 1 + 1 + 3
    ROGER: 10I, 11I, 1 + 1 + 3
    RITA: 1 + 1 + 3

The "conversation" continues along much these same lines, until Rita finally changes her tune:

    RITA: 1 + 1 + 3
    ROGER: 12R, 10I, 10I, 9I, 9I
    RITA: 9I, 8I

Not long ago, suffering from writer's block, I asked ChatGPT if it could rewrite *Moby-Dick* from the whale's perspective. The chatbot began modestly. "As an A.I. language model, I can certainly attempt to rewrite a small excerpt," it told me. Then it launched into what it titled *Moby-Dick—The White Leviathan's Tale.*

In Moby-Dick's *Moby-Dick*, the plot turned on an unfortunate case of mistaken identity. Some other whale had gnawed off Ahab's leg; the white whale was as innocent as the driven snow.

"My heart was one of peace, my spirit one with the rhythm of the currents," ChatGPT wrote:

> I knew not of the vendetta forming against me, for I was but a creature of instinct, a guardian of the seas. But Captain Ahab's thirst for vengeance loomed large, and I became the object of his undying rage. The scar he bore from a previous encounter with my kind had branded me as his ultimate adversary. In his eyes, I represented a force of nature that defied his understanding, a living enigma of the ocean's might.

In paragraph seven, I was surprised—and excited—to see the whale/bot allude to the problem of interspecies communication:

> Deep within the abyss of the sea, I pondered the futility of the captain's quest. I longed to communicate with him, to show him that my kind, too, had emotions and families to protect. But the language barrier between our worlds remained an insurmountable chasm.

As anyone who has been conscious for the past ten months knows, ChatGPT is capable of amazing feats. It can write essays, compose sonnets, explain scientific concepts, and produce jokes (though these last are not necessarily funny). If you ask ChatGPT how it was created, it will tell you that first it was trained on a "massive corpus" of data from the Internet. This phase consisted of what's called "unsupervised machine learning," which was performed by an intricate array of processing nodes known as a neural network. Basically, the "learning" involved filling in the blanks; according to ChatGPT, the exercise entailed "predicting the next word in a sentence given the context of the previous words." By digesting millions of web pages—and calculating and recalculating the odds—ChatGPT got so good at this guessing game that, without ever understanding English, it mastered the language. (Other languages it is "fluent" in include Chinese, Spanish, and French.)

In theory at least, what goes for English (and Chinese and French) also goes for sperm whale. Provided that a computer model can be trained on enough data, it should be able to master coda prediction. It could then—once again in theory—generate sequences of codas that a sperm whale would find convincing. The model wouldn't understand sperm whale–ese, but it could, in a manner of speaking, speak it. Call it ClickGPT.

Currently, the largest collection of sperm-whale codas is an archive assembled by Gero in his years on and off Dominica. The codas contain roughly a hundred thousand clicks. In a paper published last year, members of the CETI team estimated that, to fulfill its goals, the project would need to assemble some four billion clicks, which is to say, a collection roughly forty thousand times larger than Gero's.

"One of the key challenges toward the analysis of sperm whale (and more broadly, animal) communication using modern deep learning techniques is the need for sizable datasets," the team wrote.

In addition to bugging individual whales, CETI is planning to tether a series of three "listening stations" to the floor of the Caribbean Sea. The stations should be able to capture the codas of whales chatting up to twelve miles from shore. (Though inaudible above the waves, sperm-whale clicks can register up to 230 decibels, which is louder than a gunshot or a rock concert.) The information gathered by the stations will be less detailed than what the tags can provide, but it should be much more plentiful.

One afternoon, I drove with Gruber and CETI's station manager, Yaniv Aluma, a former Israeli Navy SEAL. to the port in Roseau, where pieces of the listening stations were being stored. The pieces were shaped like giant sink plugs and painted bright yellow. Gruber explained that the yellow plugs were buoys, and that the listening equipment—essentially, large collections of hydrophones—would dangle from the bottom of the buoys, on cables. The cables would be weighed down with old train wheels, which would anchor them to the seabed. A stack of wheels, rusted orange, stood nearby. Gruber suddenly turned to Aluma and, pointing to the pile, said, "You know, we're going to need more of these." Aluma nodded glumly.

The listening stations have been the source of nearly a year's worth of delays for CETI. The first was installed last summer, in water six thousand feet deep. Fish were attracted to the buoy, so the spot soon became popular among fishermen. After about a month, the fishermen noticed that the buoy was gone. Members of CETI's Dominica-based staff set out in the middle of the night on *CETI 1* to try to retrieve it. By the time they reached the buoy, it had drifted almost thirty miles offshore. Meanwhile, the hydrophone array, attached to the rusty train wheels, had dropped to the bottom of the sea.

The trouble was soon traced to the cable, which had been manufactured in Texas by a company that specializes in offshore oil-rig equipment. "They deal with infrastructure that's very solid," Aluma explained. "But a buoy has its own life. And they didn't calculate so well the torque or load on different motions—twisting and moving sideways." The company spent months figuring out why the cable had failed and finally thought it had solved the problem. In June, Aluma flew to Houston to watch a new cable go through stress tests. In the middle of the tests, the new design failed. To avoid further delays, the CETI team reconfigured the stations. One of the reconfigured units was installed late last month. If it doesn't float off, or in some other way malfunction, the plan is to get the two others in the water sometime this fall.

A sperm whale's head takes up nearly a third of its body; its narrow lower jaw seems borrowed from a different animal entirely; and its flippers are so small as to be almost dainty. (The formal name for the species is *Physeter macrocephalus*, which translates roughly as "big-headed blowhole.") "From just about any angle," Hal Whitehead, one of the world's leading sperm-whale experts (and Gero's thesis adviser), has written, sperm whales appear "very strange." I wanted to see more of these strange-looking creatures than was visible from a catamaran, and so, on my last day in Dominica, I considered going on a commercial tour that offered customers a chance to swim with whales, assuming that any could be located. In the end—partly because I sensed that Gruber disapproved of the practice—I dropped the idea.

Instead, I joined the crew on *CETI 1* for what was supposed to be another round of drone tagging. After we'd been under way for about two hours, codas were picked up, to the northeast. We headed in that direction and soon came upon an extraordinary sight. There were at least ten whales right off the boat's starboard. They were all facing the same direction, and they were bunched

tightly together, in rows. Gero identified them as members of Unit
A. The members of Unit A were originally named for characters
in Margaret Atwood novels, and they include Lady Oracle, Aurora,
and Rounder, Lady Oracle's daughter.

Earlier that day, the crew on *CETI 2* had spotted pilot whales,
or blackfish, which are known to harass sperm whales. "This looks
very defensive," Gero said, referring to the formation.

Suddenly, someone yelled out, "Red!" A burst of scarlet spread
through the water, like a great banner unfurling. No one knew
what was going on. Had the pilot whales stealthily attacked? Was
one of the whales in the group injured? The crowding increased
until the whales were practically on top of one another.

Then a new head appeared among them. "Holy fucking shit!"
Gruber exclaimed.

"Oh, my God!" Gero cried. He ran to the front of the boat,
clutching his hair in amazement. "Oh, my God! Oh, my God!"
The head belonged to a newborn calf, which was about twelve feet
long and weighed maybe a ton. In all his years of studying sperm
whales, Gero had never watched one being born. He wasn't sure
anyone ever had.

As one, the whales made a turn toward the catamaran. They
were so close I got a view of their huge, eerily faceless heads and
pink lower jaws. They seemed oblivious of the boat, which was
now in their way. One knocked into the hull, and the foredeck
shuddered.

The adults kept pushing the calf around. Its mother and her
relatives pressed in so close that the baby was almost lifted out of
the water. Gero began to wonder whether something had gone
wrong. By now, everyone, including the captain, had gathered
on the bow. Pagani and another undergraduate, Aidan Kenny,
had launched two drones and were filming the action from the
air. Mevorach, meanwhile, was recording the whales through a
hydrophone.

To everyone's relief, the baby began to swim on its own. Then
the pilot whales showed up—dozens of them.

"I don't like the way they're moving," Gruber said.

"They're going to attack for sure," Gero said. The pilot whales'
distinctive, wave-shaped fins slipped in and out of the water.

What followed was something out of a marine-mammal *Lord of
the Rings*. Several of the pilot whales stole in among the sperm

whales. All that could be seen from the boat was a great deal of thrashing around. Out of nowhere, more than forty Fraser's dolphins arrived on the scene. Had they come to participate in the melee or just to rubberneck? It was impossible to tell. They were smaller and thinner than the pilot whales (which, their name notwithstanding, are also technically dolphins).

"I have no prior knowledge upon which to predict what happens next," Gero announced. After several minutes, the pilot whales retreated. The dolphins curled through the waves. The whales remained bunched together. Calm reigned. Then the pilot whales made another run at the sperm whales. The water bubbled and churned.

"The pilot whales are just being pilot whales," Gero observed. Clearly, though, in the great "struggle for existence," everyone on board *CETI 1* was on the side of the baby.

The skirmishing continued. The pilot whales retreated, then closed in again. The drones began to run out of power. Pagani and Kenny piloted them back to the catamaran to exchange the batteries. These were so hot they had to be put in the boat's refrigerator. At one point, Gero thought that he spied the new calf, still alive and well. (He would later, from the drone footage, identify the baby's mother as Rounder.) "So that's good news," he called out.

The pilot whales hung around for more than two hours. Then, all at once, they were gone. The dolphins, too, swam off.

"There will never be a day like this again," Gero said as *CETI 1* headed back to shore.

That evening, everyone who'd been on board *CETI 1* and *CETI 2* gathered at a dockside restaurant for a dinner in honor of the new calf. Gruber made a toast. He thanked the team for all its hard work. "Let's hope we can learn the language with that baby whale," he said.

I was sitting with Gruber and Gero at the end of a long table. In between drinks, Gruber suggested that what we had witnessed might not have been an attack. The scene, he proposed, had been more like the last act of *The Lion King*, when the beasts of the jungle gather to welcome the new cub.

"Three different marine mammals came together to celebrate and protect the birth of an animal with a sixteen-month gestation period," he said. Perhaps, he hypothesized, this was a survival tactic that had evolved to protect mammalian young against sharks, which would have been attracted by so much blood and which, he pointed out, would have been much more numerous before humans began killing them off.

"You mean the baby whale was being protected by the pilot whales from the sharks that aren't here?" Gero asked. He said he didn't even know what it would mean to test such a theory. Gruber said they could look at the drone footage and see if the sperm whales had ever let the pilot whales near the newborn and, if so, how the pilot whales had responded. I couldn't tell whether he was kidding or not.

"That's a nice story," Mevorach interjected.

"I just like to throw ideas out there," Gruber said.

> "My! You don't say so!" said the Doctor. "You never talked that way to me before."
> "What would have been the good?" said Polynesia, dusting some cracker crumbs off her left wing. "You wouldn't have understood me if I had."
>
> —*The Story of Doctor Dolittle*

The Computer Science and Artificial Intelligence Laboratory (CSAIL), at M.I.T., occupies a Frank Gehry–designed building that appears perpetually on the verge of collapse. Some wings tilt at odd angles; others seem about to split in two. In the lobby of the building, there's a vending machine that sells electrical cords and another that dispenses caffeinated beverages from around the world. There's also a yellow sign of the sort you might see in front of an elementary school. It shows a figure wearing a backpack and carrying a briefcase and says "NERD XING."

Daniela Rus, who runs CSAIL (pronounced "see-sale"), is a roboticist. "There's such a crazy conversation these days about machines," she told me. We were sitting in her office, which is dominated by a robot, named Domo, who sits in a glass case. Domo has a metal torso and oversized, goggly eyes. "It's either machines are going to take us down or machines are going to solve all of our problems. And neither is correct."

Along with several other researchers at CSAIL, Rus has been thinking about how CETI might eventually push beyond coda prediction to something approaching coda comprehension. This is a formidable challenge. Whales in a unit often chatter before they dive. But what are they chattering about? How deep to go, or who should mind the calves, or something that has no analogue in human experience?

"We are trying to correlate behavior with vocalization," Rus told me. "Then we can begin to get evidence for the meaning of some of the vocalizations they make."

She took me down to her lab, where several graduate students were tinkering in a thicket of electronic equipment. In one corner was a transparent plastic tube loaded with circuitry, attached to two white plastic flippers. The setup, Rus explained, was the skeleton of a robotic turtle. Lying on the ground was the turtle's plastic shell. One of the students hit a switch and the flippers made a paddling motion. Another student brought out a two-foot-long robotic fish. Both the fish and the turtle could be configured to carry all sorts of sensors, including underwater cameras.

"We need new methods for collecting data," Rus said. "We need ways to get close to the whales, and so we've been talking a lot about putting the sea turtle or the fish in water next to the whales, so that we can image what we cannot see."

CSAIL is an enormous operation, with more than fifteen hundred staff members and students. "People here are kind of audacious," Rus said. "They really love the wild and crazy ideas that make a difference." She told me about a diver she had met who had swum with the sperm whales off Dominica and, by his account at least, had befriended one. The whale seemed to like to imitate the diver; for example, when he hung in the water vertically, it did, too.

"The question I've been asking myself is: Suppose that we set up experiments where we engage the whales in physical mimicry," Rus said. "Can we then get them to vocalize while doing a motion? So, can we get them to say, 'I'm going up'? Or can we get them to say, 'I'm hovering'? I think that, if we were to find a few snippets of vocalizations that we could associate with some meaning, that would help us get deeper into their conversational structure."

While we were talking, another CSAIL professor and CETI collaborator, Jacob Andreas, showed up. Andreas, a computer scientist who works on language processing, said that he had been introduced to the whale project at a faculty retreat. "I gave a talk about understanding neural networks as a weird translation problem," he recalled. "And Daniela came up to me afterwards and she said, 'Oh, you like weird translation problems? Here's a weird translation problem.'"

Andreas told me that CETI had already made significant strides, just by reanalyzing Gero's archive. Not only had the team uncovered the new kind of signal but also it had found that codas have much more internal structure than had previously been recognized. "The amount of information that this system can carry is much bigger," he said.

"The holy grail here—the thing that separates human language from all other animal communication systems—is what's called 'duality of patterning,'" Andreas went on. "Duality of patterning" refers to the way that meaningless units—in English, sounds like "sp" or "ot"—can be combined to form meaningful units, like "spot." If, as is suspected, clicks are empty of significance but codas refer to something, then sperm whales, too, would have arrived at duality of patterning. "Based on what we know about how the coda inventory works, I'm optimistic—though still not sure—that this is going to be something that we find in sperm whales," Andreas said.

The question of whether any species possesses a "communication system" comparable to that of humans is an open and much debated one. In the nineteen-fifties, the behaviorist B. F. Skinner argued that children learn language through positive reinforcement; therefore, other animals should be able to do the same. The linguist Noam Chomsky had a different view. He dismissed the notion that kids acquire language via conditioning, and also the possibility that language was available to other species.

In the early nineteen-seventies, a student of Skinner's, Herbert Terrace, set out to confirm his mentor's theory. Terrace, at that point a professor of psychology at Columbia, adopted a chimpanzee, whom he named, tauntingly, Nim Chimpsky. From the age of two weeks, Nim was raised by people and taught American Sign Language. Nim's interactions with his caregivers were videotaped, so that Terrace would have an objective record of the chimp's progress. By the time Nim was three years old, he had a repertoire of eighty signs and, significantly, often produced them in sequences, such as "banana me eat banana" or "tickle me Nim play." Terrace set out to write a book about how Nim had crossed the language barrier and, in so doing, made a monkey of his namesake. But then Terrace double-checked some details of his account against the tapes. When he looked carefully at the videos, he was appalled. Nim hadn't really learned ASL; he had just learned to imitate the last signs his teachers had made to him.

"The very tapes I planned to use to document Nim's ability to sign provided decisive evidence that I had vastly overestimated his linguistic competence," Terrace wrote.

Since Nim, many further efforts have been made to prove that different species—orangutans, bonobos, parrots, dolphins—have a capacity for language. Several of the animals who were the focus of these efforts—Koko the gorilla, Alex the gray parrot—became international celebrities. But most linguists still believe that the only species that possesses language is our own.

Language is "a uniquely human faculty" that is "part of the biological nature of our species," Stephen R. Anderson, a professor emeritus at Yale and a former president of the Linguistic Society of America, writes in his book *Doctor Dolittle's Delusion*.

Whether sperm-whale codas could challenge this belief is an issue that just about everyone I talked to on the CETI team said they'd rather not talk about.

"Linguists like Chomsky are very opinionated," Michael Bronstein, the Oxford professor, told me. "For a computer scientist, usually a language is some formal system, and often we talk about artificial languages." Sperm-whale codas "might not be as expressive as human language," he continued. "But I think whether to call it 'language' or not is more of a formal question."

"Ironically, it's a semantic debate about the meaning of 'language,'" Gero observed.

Of course, the advent of ChatGPT further complicates the debate. Once a set of algorithms can rewrite a novel, what counts as "linguistic competence"? And who—or what—gets to decide?

"When we say that we're going to succeed in translating whale communication, what do we mean?" Shafi Goldwasser, the Radcliffe Institute fellow who first proposed the idea that led to CETI, asked.

"Everybody's talking these days about these generative A.I. models like ChatGPT," Goldwasser, who now directs the Simons Institute for the Theory of Computing, at the University of California, Berkeley, went on. "What are they doing? You are giving them questions or prompts, and then they give you answers, and the way that they do that is by predicting how to complete sentences or what the next word would be. So you could say that's a goal for CETI—that you don't necessarily understand what the whales are saying, but that you could predict it with good success. And, therefore, you could maybe generate a conversation that would be understood by a whale, but maybe you don't understand it. So that's kind of a weird success."

Prediction, Goldwasser said, would mean "we've realized what the pattern of their speech is. It's not satisfactory, but it's something.

"What about the goal of understanding?" she added. "Even on that, I am not a pessimist."

There are now an estimated eight hundred and fifty thousand sperm whales diving the world's oceans. This is down from an estimated two million in the days before the species was commercially hunted. It's often suggested that the darkest period for *P. macrocephalus* was the middle of the nineteenth century, when Melville shipped out of New Bedford on the *Acushnet*. In fact, the bulk of the slaughter took place in the middle of the twentieth century, when sperm whales were pursued by diesel-powered ships the size of factories. In the eighteen-forties, at the height of open-boat whaling, some five thousand sperm whales were killed each year; in the nineteen-sixties, the number was six times as high. Sperm whales were boiled down to make margarine, cattle feed, and glue. As recently as the nineteen-seventies, General Motors used spermaceti in its transmission fluid.

Near the peak of industrial whaling, a biologist named Roger Payne heard a radio report that changed his life and, with it, the lives of the world's remaining cetaceans. The report noted that a whale had washed up on a beach not far from where Payne was working, at Tufts University. Payne, who'd been researching moths, drove out to see it. He was so moved by the dead animal that he switched the focus of his research. His investigations led him to a naval engineer who, while listening for Soviet submarines, had recorded eerie underwater sounds that he attributed to humpback whales. Payne spent years studying the recordings; the sounds, he decided, were so beautiful and so intricately constructed that they deserved to be called "songs." In 1970, he arranged to have *Songs of the Humpback Whale* released as an LP.

"I just thought: the world has to hear this," he would later recall. The album sold briskly, was sampled by popular musicians like Judy Collins, and helped launch the "Save the Whales" movement. In 1979, *National Geographic* issued a "flexi disc" version of the songs, which it distributed as an insert in more than ten million copies of the magazine. Three years later, the International Whaling Commission declared a "moratorium" on commercial hunts which remains in effect today. The move is credited with having rescued several species, including humpbacks and fin whales, from extinction.

Payne, who died in June at the age of eighty-eight, was an early and ardent member of the CETI team. (This was the case, Gruber told me, even though he was disappointed that the project was focusing on sperm whales, rather than on humpbacks, which, he maintained, were more intelligent.) Just a few days before his death, Payne published an op-ed piece explaining why he thought CETI was so important.

Whales, along with just about every other creature on Earth, are now facing grave new threats, he observed, among them climate change. How to motivate "ourselves and our fellow humans" to combat these threats?

"Inspiration is the key," Payne wrote. "If we could communicate with animals, ask them questions and receive answers—no matter how simple those questions and answers might turn out to be—the world might soon be moved enough to at least start the process of halting our runaway destruction of life."

Several other CETI team members made a similar point. "One important thing that I hope will be an outcome of this project has to do with how we see life on land and in the oceans," Bronstein said. "If we understand—or we have evidence, and very clear evidence in the form of language-like communication—that intelligent creatures are living there and that we are destroying them, that could change the way that we approach our Earth."

"I always look to Roger's work as a guiding star," Gruber told me. "The way that he promoted the songs and did the science led to an environmental movement that saved whale species from extinction. And he thought that CETI could be much more impactful. If we could understand what they're saying, instead of 'save the whales' it will be 'saved by the whales.'

"This project is kind of an offering," he went on. "Can technology draw us closer to nature? Can we use all this amazing tech we've invented for positive purposes?"

ChatGPT shares this hope. Or at least the A.I.-powered language model is shrewd enough to articulate it. In the version of *Moby-Dick* written by algorithms in the voice of a whale, the story ends with a somewhat ponderous but not unaffecting plea for mutuality:

I, the White Leviathan, could only wonder if there would ever come a day when man and whale would understand each other, finding harmony in the vastness of the ocean's embrace.

ISOBEL WHITCOMB

Homeward Bound

FROM *Sierra*

AMONG THE SPRAWLING wetlands and ponderosa-pine forests of the Upper Klamath Basin, fish once gave people life. In this part of Southern Oregon, winters can be long, harsh, and hungry. The Klamath Tribes tell a story that one particularly deadly winter, a spirit in the form of a giant serpent preyed upon villagers, picking people off one by one. When the villagers begged for relief from their creator, Gmok'amp'c, the god took the serpent to a mountaintop, chopped it into pieces, and threw the pieces into a lake—where they swam away as fish. The creator then announced that as long as the fish known as the c'waam and koptu remain, the Klamath people will continue to exist.

For generations, c'waam and koptu—also called Lost River suckers and shortnosed suckers—churned up the Sprague and Williamson Rivers at the end of every winter, when snow still lay on the ground and food stores grew scarce. Each year, the water writhed with fish. Weary of the winter, people native to this region, including those from the Klamath Tribe and the Modoc Nation, rushed to the rivers to fill basket after basket.

Klamath tribal member Garin Riddle Sr., also known as Big Badger, shared the origin story of the fish during a C'waam Ceremony this past April. Klamath elders and others from the community gathered around a crackling bonfire on the banks of the Sprague River to celebrate the c'waam and koptu's return to their spawning waters—a ceremony dating back thousands of years. Patchy snow still covered the ground. The river, brown and turbid with snowmelt, was decisively empty of the fish.

"We almost didn't have a C'waam Ceremony," Riddle said.

Traditionally, the ceremony involves blessing two c'waam—one released back into the river, the other placed on the fire. This year, it had taken days and multiple pairs of eyes to track down and catch a single fish, which swam in a large black bucket at the center of the crowd of onlookers. The c'waam was a funny-looking creature—two feet long with a rounded and slightly upturned nose, like a bulldog's. Its huge pink lips, which had evolved to scour lake and river bottoms for algae and phytoplankton, worked furiously. It was also beautiful, with navy blue scales and a tail fin that spread out like a folded-paper fan. "It was a struggle to catch this fish," Riddle said.

The suckers are teetering toward extinction. Each year, fewer than 30,000 c'waam and 6,500 koptu migrate back to their spawning grounds, down from the tens of millions that once thrived in the Upper Klamath Basin. Decades of habitat degradation and pollution from agriculture have rendered Upper Klamath Lake, the home to which they return after they finish spawning, poisonous. While most adult fish can withstand the pollution, it's too toxic for their young to survive. In 1988, the fish were classified as endangered species. Not a single juvenile has lived past infancy since the 1990s, and the hardy adult survivors are steadily growing older. Each year, the c'waam and koptu make their run, spawn, and lay eggs. And each year, their populations shrink.

Decades of legal battles and a new captive-rearing program have kept the c'waam and koptu on this earth. A large-scale reintroduction effort aims to return these captive-raised fish to the home that's in their DNA. For the Klamath Tribes, it's a fight for culture, community, and the right to exist as they have for time immemorial.

At the ceremony, the wild c'waam was passed around to the elders for them to touch and pray over it. Its eyes darted wildly as the elders put their hands on its gaping mouth, like a kiss. "The prophecy of this fish going away—if that happens, we will follow," Riddle said. "Now we are tasked to go ahead and honor that fish."

More than 150 years ago, a complex of wetlands filled the Upper Klamath Basin. Vast plains of water stretched all the way to the mountains on the horizon. Rivers snaked haphazardly, sprawling into bogs before re-forming. Tule reed formed golden fields, interspersed with bright-yellow water lilies called wocus. This lush vegetation provided shelter to the c'waam and koptu for the first two to five years of their life cycle, when the young fish are too vulnerable to survive in the open waters of Upper Klamath Lake. They enjoyed long life spans, surviving well into their forties and fifties.

This was also the environment where the Klamath Tribes thrived before colonization. Native people, living in small settlements across the region, collected the tule and wocus, hunted for deer and elk in the ponderosa forests, and fished for c'waam and koptu. Then, in the mid-1800s, white settlers began trickling into the Upper Klamath Basin, lured by gold and land. Already, the landscape had been altered by fur trappers, who had decimated the engineers of the lush wetlands: beavers. The new settlers further transformed it, draining the swamps and lakes to farm the fertile peat soil.

With many of the wetlands gone or fenced off, it became nearly impossible for Native tribes to live as they once had. Many were starving. To survive, the Klamath Tribes were forced to make a compromise: In the Treaty of 1864, leaders ceded 23 million acres of land to the US government in exchange for the right to hunt, fish, and gather on their remaining 2 million acres "in perpetuity." Soldiers from the newly built Fort Klamath rounded up the tribes and placed them in a single, cramped reservation with very few resources. Leaders were publicly executed and traditional religious practices outlawed.

Over the following decades, the US government handed out parcels of the Klamath Tribes' land for pennies—to early settlers and then World War I veterans. Wetlands continued to disappear, replaced by fields of alfalfa and potatoes. Without beavers, the Sprague and Williamson Rivers turned into narrow channels. After the Bureau of Reclamation installed a network of dams and irrigation canals, known as the Klamath Project, nutrient pollution flowed unfiltered from the farmlands into Upper Klamath Lake.

As the suckers' habitat deteriorated, so did the tribes' traditional way of life—including their relationship with the c'waam and koptu. In 1932, the Klamath Tribes stopped gathering for the C'waam Ceremony—the beginning of a sixty-year-long lapse in the ritual. No one alive today knows why. But Jeff Mitchell, a tribal council member who stood at the 2023 ceremony to describe the last time the celebration was practiced "in the old ways," has a theory: grief.

"The man who led the last ceremony, Sky, he walked this land when it was just us here," Mitchell said. "Before our neighbors arrived, before we signed a treaty with the US. Told we can't speak our language no more, told we can't fish and hunt. He had seen so much change."

The Klamath people stopped preparing c'waam and koptu as they once did—opting to batter and fry the fish instead of hanging it to dry. Children were sent to residential schools, where abuse was rampant and they were barred from speaking their own language. Then, in 1954, Congress enacted the Termination Act, which stripped the Klamath Tribes of their tribal status and the federal protections that came with it. As a result, they lost all their land. "[People] scattered. Alcoholism rates soared. Suicide rates soared," said Clay Dumont, chairman of the Klamath Tribes.

By 1988, when the c'waam and koptu were first listed as endangered, the richness of all that the Klamath Tribes had before colonization—their hunting, gathering, and fishing way of life—was in danger of being lost. The community was fractured. The loss of the C'waam Ceremony had been particularly devastating to tribal identity: "[The ceremonies] are important because they produce a unity of understanding," Dumont said. "You lose those shared understandings, shared respect for who we are and how we're attached to a place."

Still, memories of the c'waam rituals hadn't completely faded. Motivated by the federal government's endangered species listing, cultural scholar and tribal member Gordon Bettles proposed an idea to tribal elders: Bring the ceremony back.

Their response wasn't an immediate yes. No one remembered exactly how the ceremony was conducted or the minutiae of the rituals; after all, the last surviving tribal members had been children when they'd last attended one. They'd need to start over. For a few years, leaders deliberated. Their decision: They would do their best with what they knew.

The return of the C'waam Ceremony in 1990 was part of a larger cultural resurgence within the Klamath Tribes and across North American Indigenous communities. Around the same time the c'waam and koptu were listed as endangered, the Klamath Tribes regained their tribal status. The change didn't return their land, but it restored access to government benefits and programs, and, most important, it restored the Klamath Tribes' sovereignty, allowing them to reassemble their tribal councils and Indigenous leadership. "This was slow at first," said Boyd Cothran, a historian focusing on Indigenous peoples and settler colonialism in the American West.

In the Klamath region and across the country, new gas stations and casinos went up, boosting tribal economies. Leaders and activists worked to keep Indigenous languages alive and fought for Native rights to water and other resources. Bettles was one of those leaders. In the 1990s, he helped launch a Klamath-Modoc language program and develop the *Klamath Words and Phrases Book*, and he taught the very first language class.

Then, in 2005, a landmark court case affirmed the tribes' water rights. One particularly dry summer several years earlier, the Bureau of Reclamation had shut down the Klamath Project to preserve minimum lake levels and streamflows for suckers and salmon. Irrigators sued the federal agency, but the court ruled in favor of the tribes, deciding that the suckers' status as endangered and Klamath water rights took priority. The case affirmed that these rights dated back to "time immemorial." "That's earlier than anybody's," said Adell Amos, a legal scholar at the University of Oregon whose work focuses on water law. Since then, the Klamath Tribes have been engaged in multiple legal fights to enforce the decision.

The Klamath Tribes' movement to reclaim their cultural heritage continues. That first generation of kids after the tribes regained their sovereignty are adults now. Many are learning to speak the Klamath language. Some, like Dumont, studied Native issues and became active in cultural groups. "Cultural revitalization is hard work. But the people are hungry for it," Dumont said.

And despite the transformation of their habitat, the remaining suckers do return year after year to the places where they first spawned. One of those places is Sucker Springs, a source of magma-warmed water that pours into Upper Klamath Lake from its eastern shores. Year-round, the water temperature at Sucker Springs remains a balmy 60.8°F, the perfect environment for embryonic fish to develop. While most c'waam and koptu migrate upriver, around 4,000 spawn at Sucker Springs.

Cued by warming springtime temperatures, the fish gather at Modoc Point before swimming together to Sucker Springs. "We really have no idea how they know to go to that particular place," said Alex Gonyaw, head biologist for the Klamath Tribes. "Just wild stuff."

The day after the April C'waam Ceremony, bright sunlight poured down on the crystalline waters of Upper Klamath Lake. In the distance, snowcapped, deep-blue mountains and sentinel-like Mt. Shasta towered over the landscape. "It's so pretty out. But it's deceptive," Gonyaw said.

Most summers, the lake turns green with mats of toxic cyano-bacteria, fueled by nutrient pollution. By fall, the cyanobacteria die, and their decomposing bodies suck oxygen from the lake's bottom. During the worst years, Gonyaw finds fish washed up dead on the shore, unable to breathe. "In September, it's a mess. It smells like a sewer out here," he said.

Gonyaw isn't a Klamath tribal member. He had never heard of the c'waam or koptu until he moved to Klamath Falls, just south of the Klamath reservation, eighteen years ago. Since he began working for the Klamath Tribes, he has come to know the suckers intimately. He's enchanted by their deep roots in this place, which go back 3 million years, when the suckers swam west across rivers that once connected the Pacific to the Mississippi basin—before the Rocky Mountains thrust upward, separating the c'waam and koptu from their mid-western brethren. "When you catch a c'waam, it's like picking up a dinosaur," he said. "They evolved here and only exist here."

Since 2016, Gonyaw has spent nearly every day out on the lake in his skiff. Even in that short time, he's watched conditions deteriorate. When he started, five times as many suckers were spawning. Each year, that number dropped by roughly 15 percent. Then, in 2022, no c'waam or koptu were detected spawning in the Williamson River. The water in the lake was too low for them to access the river's mouth. "It was very disturbing," Gonyaw said.

The weariness of bearing witness to this decline shows on his face. Gonyaw's warm brown eyes have heavy bags beneath them, the corners of his mouth turn down, and his brow is nearly always furrowed. "It's been constantly watching things disappear," he said. "This relentless chipping away of everything."

The c'waam and koptu are knitted into the fabric of this ecosystem. They harvest energy from microscopic lake species and use that energy to feed an entire food web—from the bald and golden eagles that still crowd the trees around Sucker Springs to the trees and wetlands fertilized by decaying sucker bodies. "The more I work with living things, there seems to be a magical component," he said.

When Gonyaw first started working with the Klamath Tribes, they were hesitant to interfere with that magic. There was talk of launching a captive-rearing program to bolster the dwindling populations, but tribal leaders were apprehensive because of the legacy of salmon hatcheries. Historically, those facilities, which breed and raise juveniles before sending them on their way, were set up in part to uphold Native treaty rights to salmon. Although they're intended to give salmon populations a boost, they have contributed to their decline: Hatchery salmon breed with wild-born fish, weakening their resilience, and they increase competition for food and other resources. "[The salmon hatcheries] didn't really do what they said they were going to do—this idea that you can build a dam, put a hatchery below the dam, and compensate for all of nature upstream," Gonyaw said. "So we avoided that until the very end. You could argue either way whether that was a good decision or not."

Then, one brutal summer forced a decision. In 2017, for the fifth year in a row, the Bureau of Reclamation responded to worsening drought by releasing more lake water into the Klamath Project than the maximum mandated under the Endangered Species Act. The lake levels got too low to allow the c'waam and koptu to access key habitat—from spawning waters to foraging sites. The low water levels had also concentrated agricultural pollution into a slurry where algae grew amok. That particular year, the pollution became too much for the fish. By the end of the summer, the Klamath Tribes had collected over 700 dead adult c'waam and eight koptu from the shores of the lake.

"That may have been a tipping point," Gonyaw said. Sucker mortality rates have steadily increased ever since.

The massive fish kill forced the tribes to set aside their hesitations about a captive-rearing program. In 2018, for the first time, Gonyaw and his colleagues scooped up buckets of c'waam and koptu eggs at Sucker Springs. Then they embarked on an experiment.

The Klamath Tribes weren't the first group to raise suckers in captivity—the U.S. Fish and Wildlife Service had begun a similar program several years prior—but the tribes began writing their own protocol as they went. When Gonyaw had scoured the literature earlier, he'd found plenty of information on raising salmon and trout, but nothing on how to raise suckers.

Newly hatched c'waam and koptu don't look much like fish. As the young fish grow, their food needs change constantly. "They evolved in a wetland environment with a banquet of food choices," Gonyaw said. He tried feeding them plankton and manufactured feed but finally settled on brine shrimp—not what suckers eat in the wild, but the ravenous babies seemed to like it just fine. The catch to that: The shrimp needed to be hatched daily in salt water in a temperature-regulated environment.

At the hatchery, in a tub the size of a kiddie pool swam a cloud of inch-long threads, each with two tiny eyeballs: baby suckers. Fishery technician and Klamath tribal member Charlie Wright bent over the tub. The fish darted away from her shadow in a swarm. "We understand it's a life, not a paycheck," Wright said. She wore her long, dark-brown hair in two braids tied together in a ponytail. A tattoo of a crucifix emerged behind her ear; her forearms, decorated with roses, were muscular from cage fighting. Since 2022, Wright has worked with Gonyaw to raise these fish.

For Wright, a typical day involves checking the ponds—their temperature and oxygen levels—and the eggs, which bounce around in oxygenated glass cylinders. Then she moves hatched babies into aquariums. Like human infants, the young fish can't go more than a few hours without eating before they become unhappy, so work is around the clock. Sometimes Wright drives up the winding mountain road to the hatchery in the middle of the night just to check on them. "There are nights you don't get any sleep," she said.

Wright grew up in the Upper Klamath Basin. "I've been hunting and fishing, practicing our treaty rights since I was a little girl out with my dad," she said. "I got to see firsthand how much things have changed." She was aware of the existence of the c'waam and koptu, and of their importance, but she had no idea how close they were to the brink of extinction until four years ago, when she began fighting for her treaty rights.

When Klamath leaders signed the Treaty of 1864, which surrendered nearly all the Klamath homelands to the US government, it was upon the condition that they'd retain the right to hunt and fish as they always had, on the little land they still had access to.

Without the ability to fish for c'waam and koptu, Klamath leaders and local Indigenous rights activists argue, those treaty rights are unfulfilled. And this loss isn't a faultless tragedy, they maintain—at least in part, it can be attributed to the reckless allocation of lake water to agriculture.

Wright's activism began on the trips she took hunting and foraging with her three young sons. Out in the backcountry, near her family's land, she'd see people illegally flooding fields and pumping water from streams, letting cattle stand hocks-deep in the rivers. She started using her smartphone to snap photos and post them online. As her online activism gained traction, she began taking it to the streets, leading rallies for the c'waam and koptu.

Whenever Wright speaks about the c'waam and koptu, her voice cracks and tears fall from her eyes. After rallies, audience members often approach her, crying themselves. "They tell me, 'Wow, we had no idea,'" Wright said.

At the hatchery, it was just her and the fish—all 56,000 of them. Wright lifted a lid on another tub that contained older juveniles, three years old, each six inches in length. The fish had evolved to live in the deep, muddy reaches of the lake; they fled from the daylight as it hit them. These juveniles are in the process of being tagged, which will allow the Klamath Tribes to track their movements after releasing them into the wild. The goal: to release 10,000 fish before 2026.

Over time, Wright has become attached to the suckers, especially the youngest, sometimes talking to them as though they are her own kids. "They're our babies," she said. "I tell them I'm happy to see them go, but I hope they come back."

In recent years, the Klamath Tribes' mood has shifted from total grief to cautious hope. There's finally a reason to believe it's not too late for the c'waam and koptu.

Interior Secretary Deb Haaland has been receptive to the Klamath Tribes' efforts to protect their treaty resources, providing valuable consultation and even visiting restoration projects on the lake. The effects of climate change on the basin, from drought to wildfires, have become apparent enough that farmers recognize that "this isn't business as usual," Dumont said. A large-scale restoration project of wetlands north of Upper Klamath Lake promises to help filter nutrient pollution from agriculture and eventually provide habitat to juvenile suckers. And while for years the lake was maintained at levels lower than those mandated under the Endangered Species Act (due to lax government oversight and agriculture), thanks to court cases by the Klamath Tribes, the government changed how it manages the lake, which finally contains enough water to allow the c'waam and koptu access to their habitat. Just downriver from the lake, the largest dam removal in history is taking place—after the project is complete, it will no longer be necessary to release large "flushing flows" from the lake to wash toxic parasites from the dams.

"Big social movements take time to develop. They take generations," Cothran, the historian, noted.

On a spring day in 2022, not long after the c'waam and koptu made their annual run, fifty kindergarteners gathered on the banks of the Sprague River alongside Gonyaw. Near the small crowd was a collection of red buckets. Each contained a squirming c'waam. One by one, the students selected a bucket, carried it to the rushing water, and turned it upside down. The fish leaped into the water, returning to a home they'd never known.

The fish were part of the first cohort of captive-raised c'waam released into Upper Klamath Lake—500 total. Many of the students, about half of whom were members of the Klamath Tribes, were learning about the c'waam and koptu for the first time. "The bulk of young people don't know these things exist," Gonyaw said. "To be able to reconnect with their fish is powerful. It was wonderful to see the smiles on their faces."

Today, those fish are dispersed throughout Upper Klamath Lake. Occasionally, Gonyaw will catch sight of one, but it's hard to know how they're doing on the vast lake bottom. In spring 2024, the first of these fish should make their return migration. Some will gather at Modoc Point. Others will surge up the Sprague and Williamson Rivers. Then, the Klamath Tribes will learn if that first effort was successful. "I'm excited to know if they can come home," Wright said.

The goal is for these reintroduced fish to lay eggs that hatch and survive, breathing life into the c'waam population. But it could be years before the lake levels are consistently high enough, and the water clear enough, for that to happen. Until that time, the Klamath Tribes will do everything in their power to keep the fish on life support. "When the snow melts in the spring, they will return," Cothran said. "The Klamath people will be there too."

In the origin story of the c'waam and koptu, the creator god told the Klamath Tribes that their existence depends on the survival of these fish. In reality, it's the reverse. As long as the Klamath people survive, the c'waam and koptu will too. And the Klamath people aren't going anywhere.

The Lonely Battle to Save Species on a Tiny Speck in the Pacific

FROM *Smithsonian*

IN MAY OF 2021, Brittany Clemans and Lindsey Bull, two sea turtle biologists in their twenties, were walking around Tern Island, an incredibly remote block of land in the middle of the Pacific Ocean, when they came across a Hawaiian green sea turtle. She had crawled onto the island the night before to nest and wandered into a hole in a metal wall, likely on her way back to the water. Her front end had made it through, but the widest part of her shell got wedged in. She couldn't back up, and she'd flailed her flippers so hard trying to move forward that the rusting steel had scratched the sides of her carapace. She was lethargic. The afternoon heat threatened her life.

The two scientists were at the heart of the largest protected area in the United States, Papahānaumokuākea Marine National Monument in Northwestern Hawaii. The monument's 583,000 square miles are filled with reefs and atolls, and Tern Island is at the northern edge of an atoll called Lalo, which has a crescent-shaped reef with a curve of about 20 miles. Like other islands in the area, Tern used to shape-shift with the storms and tides, and birds, seals and turtles easily moved around its sloping shores. But in the 1940s, the Navy turned Tern into a pit stop for planes flying between Hawaii and Midway Atoll. It built the island into the form of an aircraft carrier, dredging more than 55,000 dump trucks' worth of coral from the shallows, flattening it into a runway about a half-mile long and 350 feet wide, rimming most of it with a sea wall.

That sea wall became an enormous hazard for the island's wild-life. Almost eighty years of storms have now rusted and wracked it into jagged spires and open holes, so that portions look like a witch's fingers or like Swiss cheese. Animals swim, fly or crawl through cuts or holes, and are often unable to escape. Other en-trapment hazards lurk, including old buildings that are falling apart and concrete structures that are cracking open. The Navy and then the Coast Guard occupied Tern Island until 1979, and the Coast Guard and Air Force left discarded batteries and electri-cal equipment leaking toxic contaminants.

Until about a decade ago, the U.S. Fish and Wildlife Service (USFWS) had a permanent field station on Tern Island, with groups of scientists studying and rescuing its seabirds, turtles and seals all year round. But a 2012 storm damaged the housing and operations facilities.

From that point on, a skeleton crew of scientists has been ven-turing to the island to study sea turtles and seals during field seasons that sometimes stretch from late spring to early fall.

More than 300,000 seabirds of eighteen species make their homes on Tern and other nearby islands. Critically endangered Hawaiian monk seals give birth on the shores. Sharks and fish of every color swim in the shallows amid corals the size of La-Z-Boys and kitchen tables. More than 90 percent of the sea turtles in the Hawaiian Archipelago, which stretches for roughly 1,500 miles, nest on the atoll.

The chance to spend time on Tern is exhilarating. But the work is exhausting. Every night that field season, Clemans and Bull sur-veyed the island from roughly 9 p.m. to 7 a.m., crossing back and forth across the soft sand—walk, crawl, squat, bend, think, crouch, walk. The biologists labored in the dark, as that's when the turtles emerged from the surf and crawled up on land to lay their eggs. Trekking roughly 11 miles a night, they looked for the pregnant female turtles and then numbered them, tagged them and mea-sured them. In the afternoon, they walked around the island once more looking for animals in danger.

When Clemans and Bull found the trapped female sea turtle that afternoon, they moved carefully. The animal could injure them with a powerful flap from her front flippers or by landing on their feet; she likely weighed 200 pounds or more. If physical harm came to one of them, a boat rescue was at least a few days away. They lifted the turtle onto her right side and scooted her forward until she was able to crawl to the water. The biologists felt relief, but concern. "She slowly swam away, and I do remember we discussed, 'OK, there is a possibility we might [later] find her washed up,'" Clemans later tells me. "'She might die.'"

Two hours later, they found another turtle flipped on her back beneath a rusted sea wall that stood a foot or more above the beach. Seeing the ocean so close, the turtle had likely taken a chance and crawled over the edge, nose-diving into the sand and then flipping on her carapace. "I knew immediately she was dead," says Clemans. "There was no movement. Flies were everywhere."

"You could see indents in the sand where her flippers had been trying to flip herself over," adds Bull. "And she just couldn't flip."

The researchers were sweaty and worn out. They had only slept four or five hours that day in their hot tents after patrolling the beaches the entire night, but they performed a necropsy, the animal equivalent of an autopsy. They determined the turtle was a healthy female, full of eggs and ready to nest. She had likely swum hundreds of miles to this atoll, the place of her birth, to bury her eggs. "Just finding her dead after getting that far, which is so, so discouraging," says Bull, "it definitely beat down on our team morale."

Three days later, the turtle they had rescued earlier that afternoon washed up dead, too.

By July, Clemans and Bull were clocking eighty hours of work a week. One morning, Bull came back from a survey after working sixteen or seventeen hours the day and night prior. She walked into the office tent, put her backpack down, turned the light on, and then tripped over the strap of her pack. She stumbled down on her knees, and then her body just gave out. She fell face-forward and clipped the side of her head on a metal chair before falling on her back and hitting her head on the plywood floor. On top of her sleep deprivation, now a concussion.

It took a research ship three days to get to her. During that time, the other biologists woke her up, asked her questions and carried out reflex tests with her hands. Once on the boat, it took another three days to get her to the island of Kauai for medical treatment. She told the doctor about the job, how little she slept and for how long. He responded, "That'll drive you crazy."

I know the truth to that statement. Right after working on the islands of Lalo, I lost my mind.

When I first told folks in 2003 that I was headed out to a small, remote atoll to study sea turtles, some wondered what would make me want to do such a thing.

For starters, I grew up with nine siblings and two parents, mostly in a three-bedroom house in Winona, Minnesota. My parents had one room, my brother Frank and I had another, and everybody else younger was in the third room. In high school, right before Frank went off to college, my dad said, "Congratulations, you'll have your own room." Which was weird, since he wasn't usually the congratulatory type. Soon after, I found a single bed shoved into the downstairs room we used as a sort of pantry with a thin curtain hung over the doorway. My new bedroom had a refrigerator, canned and dry goods, and—in a sign of how square I was in high school—my dad's liquor stash that didn't even tempt me.

When we weren't crammed at home, we piled ourselves into a van for road trips. We most often visited the ocean, camping on Padre Island, Texas, or crashing in hotels in Myrtle Beach, South Carolina. Highlights included seeing dolphins swimming in the surf. Once, in grade school, Frank and I came across a small shark caught by a fisherman with a rod and reel. We carried it into the ocean and held it while walking forward until it swam off.

My family played football and Frisbee, but in quieter moments, I would go alone to the beach and sculpt large sand models of dolphins, sea turtles and sharks. I grew to appreciate nature, but a key moment pushed me toward the value of science. In middle school, when I saw some friends biking behind a truck that was spraying our town for mosquitos, I ran out to play with them in the mist. My mom yelled at me to stop and told me the spray was poison. The next day she gave me Rachel Carson's *Silent Spring*, the 1962 book that famously chronicled the dangers of DDT and other pesticides, forever changing the environmental movement. My mom knew I would slough off her warning, but giving me a story with evidence would ensure that I never ran after such a truck again.

So a lot of reasons led me to take a job roughly 450 miles from civilization to study Hawaiian green sea turtles for the USFWS and National Oceanic and Atmospheric Administration (NOAA). A love of science. A longing for privacy. A sense of adventure, particularly tied to the ocean. Though I had previous experience as a herpetologist in the Caribbean and California, this would be my most isolated stint as a scientist.

I arrived on Tern Island in May 2003, with another biologist on a three-hour flight in a small plane over the open ocean. As the runway appeared below, the pilots put on their helmets. Surprised, I looked to my colleague, who explained it was in case birds crashed through the windshield. Upon making our descent, the dark expanse covering the island rose up and spread out; more than a hundred thousand birds amassed into a dark cloud. The plane yawed left and right, dipped and climbed to avoid the flocks, churning my stomach like a roller coaster. Upon landing, the deafening cries of the birds weren't even the strongest offense to the senses; the rancid smell of guano hung thick in the air.

On Tern, I spent time learning from, and sometimes helping, the seabird and seal biologists, but I primarily focused on the nesting turtles on nearby East Island, an 11-acre bump of dead coral and sand covered in a handful of bushes and some low-lying vegetation. I counted the mothers so authorities knew how many nested every year, tagged them so they could be identified when they nested again and took photos to document any notable injuries or afflictions. (Sea turtles sometimes grow tumors, more often when swimming in bays with excess nutrients and little ocean mixing.) I lived out of a tent and traded shifts with another biologist every three to seven days or so. When I traveled back to Tern, I crunched my data and slept in housing that was walled off from birds and cooler than my tent. My duties on East were much the same as Clemans and Bull's on Tern, and they usually lasted from early evening to midmorning the following day. A daily journal, which I still have and referenced for this article, chronicled the ups and downs of the routine. Looking back upon my writing two decades later, I'm transported back to how thrilled I was to snorkel with manta rays with wingspans that dwarfed mine and how startled I was when a wedge-tailed shearwater flew into my shoulder, fell to the ground and leisurely waddled off.

My first night on the island was May 22, 2003—eighteen years to the day before Bull and Clemans found that dead turtle. I had studied the reptiles before, in the U.S. Virgin Islands, but became much more adept at deciphering what stage of nesting they were in. When nesting, female turtles make sand pits a foot or two deep, then carefully scoop sand with their hind flippers to make chambers for their eggs. Once a turtle lays about 100 eggs, she pats sand gently over them and then throws sand with her front flippers to disguise the nest under a mound. I often witnessed mothers showing incredible tenacity; some had entire flippers bitten off by tiger sharks, and others exhibited fresh wounds with bones extending out. But the turtles didn't let the injuries stop them from digging, sometimes all night.

My work required tenacity, too. I walked through so much soft sand over the summer and squatted and crouched and crawled so much that I lost close to 20 pounds, despite treating every day like a carb party. The duties often ran straight through the night, so I took short food breaks, drinking cold cans of soup and chili. I made coffee doused with chocolate syrup and ate packages of Chips Ahoy! and Oreos.

Life and death packed every day. Fledgling birds flew off the island at dawn. Sometimes, they dropped in the surf just feet away from where I stood, and ten-foot-long tiger sharks jumped out of the shallows to eat them. When sea turtles crawled ashore at night to dig holes and lay their eggs, they might crush the eggs of ground-nesting seabirds or collapse the homes of burrow-nesting seabirds. Male seals cruised along the shoreline for mates. Females gave birth to their pups on the beach, leaving their placentas in the sand. Blood and sex and splattered yolks were the norm.

"The trash here is crazy," reported my journal entry from my first night on East. Washed-up nets, bottles, buoys, fishing line and broken, weathered, colorful bits of plastic lined the shore. The plastic nets and line posed a significant threat to the turtles and seals, who could easily become entangled in them and die. Since 1982, scientists have documented more than 300 entanglement incidents of Hawaiian monk seals, of which only 1,500 or so remain in the wild. Many more instances likely go undetected. Globally, at least 354 different species have been found in similarly entangled states.

On my dusk walks around the island, I played my part to make a dent, collecting debris and piling it near my camp for later removal. But the trash didn't just wash ashore. It came by air, too. Once, while I was walking on Tern's runway with a seabird biologist, we came across a dark brown mass of crud about the size of a child's drinking cup. Frans Juola, the researcher, explained that black-footed and Laysan albatross chicks, seabirds with wingspans stretching more than six feet, sometimes regurgitated their stomach contents to get rid of undigestible bits.

As Juola looked at this bolus, as the mass is called, he found bits of plastic and a roughly three-inch-long hook, thick as a chick's neck, with a metal line attached. A parent had swallowed the hook while at sea, then flown home to this island and regurgitated the sharp object to its chick, along with the debris. The chick, in turn, had coughed it up. Not all chicks were able to rid their stomachs of foreign objects. They often died, and amid their bones were the large, colorful clumps of weathered plastic—lighters, bottle caps, fishing line—that filled their stomachs. Scientists studying the chicks in Hawaii in the 1980s found that 90 percent of them already had plastics in their guts. The problem has gotten more severe, and scientists estimate that 99 percent of all seabird species will ingest plastics by 2050.

Though the plastic brought me down, so many natural events surprised me that first year. When snorkeling off Tern and around the refuge, I spied large, three-foot-long fish called jacks; whitetip reef sharks; eels; and, among the corals, loads of small, colorful creatures, from squirrelfish to nudibranchs. On East, a molting monk seal smelled worse than sweaty, pungent gym socks. A brown noddy landed on my head and performed a tap dance. Sea turtles threw sand into any exposed crevice in my body. Floating albatross feathers, the aftermath of attacks by sharks, stuck to my skin as I emerged from the ocean after a bath. Slimy guano bombs from birds smacked my head and back.

Both East and Tern were packed with wildlife that migrated away in summer and fall. On East, the turtles dug so many nests that by the end of the summer the island looked like a mogul run. Their hatchlings erupted out of the sand by the hundreds at night. On Tern, tens of thousands of seabird chicks sat less than a foot apart from each other on the ground, just waiting to lose their down and take to the skies.

By the end of the season, my sugar-fueled work had helped identify more than 200 nesting turtles, a big uptick from the sixty-seven that had been found nesting on the island in 1973. The government's 1978 listing of the green sea turtle under the Endangered Species Act had helped, though the animals were far from recovered and retained their "threatened species" status. When I had time, I helped turtles whose hind flippers had been bitten off by sharks by digging nests for them. Turtles became entangled in copper wire left in the ground by the Coast Guard, who had used East as a long-range navigation station in the 1940s and 1950s. I cut the animals free and pulled up the metal.

*

After spending four consecutive research seasons on Tern and East Islands, I was away from 2007 to 2012, having taken a job at *Outside* magazine before writing as a freelance journalist. In July 2013, I made two brief pit stops by ship to Lalo to check sea turtle nesting on its islands.

Though Tern had been falling apart since the 1970s, many damages had been mitigated by USFWS staff. But in December 2012, a powerful storm hit the island. Damaging winds, possibly over 100 miles per hour, destroyed the boat shed and tore the walls off the barracks before generating enough flying debris to kill or injure more than 200 birds, of which many had to be euthanized. Nine days after the storm, the staff were rescued. Tern would not host scientists year-round again. Dedicated biological surveys that had spanned decades stopped. Buildings fell apart. Walks to look for entrapped animals ended.

I spent just enough time on Tern the next summer to see the remains of the barracks. Rooms with walls ripped off were exposed to the ocean. Seabirds perched and nested on exposed metal frames, hanging wires and shelves where, one biologist pointed out, books about seabirds used to sit.

The next year, in 2014, I returned to study sea turtles again. The destruction of the barracks meant USFWS had nowhere to house researchers for year-round surveys, and damage to the runway made it more difficult to quickly evacuate the island before powerful storms. NOAA still funded tent-based seasonal surveys of sea turtles and seals in warmer months.

The morning we arrived, field-station manager Meg Duhr and I walked the island. Half the runway was covered in vegetation. We began a survey at the northwest end of Tern and came to a section called the Bulky Dump. Beginning in at least the 1970s, the sea wall here had started to fail, and the Coast Guard threw down debris as a defense against the ocean. Inside the rusted and bent sea wall, broken-up chunks of concrete, wires and all sorts of mechanical equipment were piled up. Water intruded into the mess. Seabirds, such as brown noddies and frigatebirds, perched on the concrete and metal. Fish darted in and out.

Duhr pointed. "There's an old propane tank," she said. "There's an old generator part." Environmental Protection Agency scientists and engineers I'd talked to before coming out there suspected the debris leaked lead and PCBs into the environment. They wanted to conduct more advanced sampling and monitoring of Tern under Superfund authorities, which can result in a designation that forces the parties responsible to clean up the contamination or reimburse the EPA for such work. They were worried the contaminants might be combining with the microplastics on the beach and in the water—and making their way into the guts of creatures that unwittingly ate them.

We moved along the north end of the island, where the ocean had rusted the sea wall into spires. Duhr said in the winter, when no one was on the island, powerful waves would sometimes push juvenile turtles through cracks between the spires and onto the sand. Sometimes they crawled back to the water, but we saw the skeleton of a juvenile drying in the middle of the island.

I kept the same nightly schedule and logged data before surveying for trapped animals in the morning. On one walk, I found a female sea turtle trapped in a hole in the sea wall. With lead monk seal biologist Shawn Farry and a few other scientists, I created temporary barriers to prevent entrapments, but waves busted down some, and sand piled up next to others, allowing turtles to crawl over them and get stuck. Later that season, I found a turtle on her back who had crawled off the edge of the sea wall, nose-dived into the sand and flipped, thrashing her flippers, one of which was bloody and injured from the fall. I crouched beside her and turned her over. She crawled into the surf and swam away.

When the sea turtle hatchlings emerged from their nests, some, rather than crawl toward the moonlight reflecting off the ocean, scampered toward the remaining bright white patches of the runway, where they would dry out and die. We all took turns collecting the youngsters in the morning and releasing them on the beach. We collected dozens if not hundreds during the field season, but once we left in September, no one would be on Tern to help them out. We listed all of the entrapments and sent the information to be shared between NOAA and USFWS authorities in Honolulu.

My favorite seabirds on the island were frigates—with their seven-foot wingspans, they are amazing aerial acrobats. One flew up and screeched at me every time I approached within ten feet of his roosting bush. On a different atoll, I once watched a frigate land on a biologist's head to rest. One time, on Tern, a male frigate swooped down and snatched expensive polarized sunglasses off of my face with its beak, soared out over the ocean, flew back over a bush packed with roosting seabirds and dropped the specs in the middle of them.

I often found frigate birds in the sea wall gap. Sometimes I climbed down to get them out. In August, I tried to lift one out while holding onto the rusted fence with one arm. Part of the sea wall broke, and I fell into the metal debris. I lifted the bird out and put it carefully down on crushed coral. A day later, I found it dead in the same spot.

Working on the islands was thrilling as always, but more exhausting than ever. In my earlier stays, I would exchange regular shifts with another biologist, traveling back and forth from Tern to East, where we each labored alone, splitting up the task of tagging sea turtles. But in 2014, I did the sea turtle surveys alone all season and spent up to fourteen straight nights all by myself on East. And when I returned to Tern, the nights of sleeping in cool, walled-off bedrooms were no more, as the buildings had been destroyed. On a satellite call, my boss and mentor George Balazs, who had worked on East beginning in the 1970s, warned me not to push myself too hard.

A short time after my arrival, I witnessed more than 470 turtles basking on the shore—most of them female, a sign that hundreds more would nest than I had ever seen before. Several nights on East saw more than 100 turtles, and I moved swiftly to cite their behavior in my notebook and measure and tag them. The demands—walk, crawl, squat, bend, think, crouch, walk—wore me down. I often noticed lightning storms on the horizon, alighting the clouds like bulbs in the color of grape sherbet. Pretty, yes, but I was the tallest thing on the island, and carrying long metal calipers.

When needed, I broke from my routine to help trapped turtles. One turtle dug herself a pit where a fishing net had been buried. Her efforts to throw sand left her entangled in debris. The plastic was wrapped around one front and one hind flipper, and left her struggling. I cut and removed the five-foot section of net from around her flippers and carried it back to camp.

In my journal, I noted that I was sleeping two to five hours a day on East, after my all-night shift had ended. I mostly dozed under a canopy tent, which offered coverage from the sun but let the trade winds cool me down during the heat of the day. Some days, the tarp on the canopy tent would come untied, flap loudly in the wind and wake me. Chalking that up to poor ropemanship skills, I kept practicing my knots, but the tarp issue continued unabated. Then, one day, I woke up around noon and saw, just past the cot, three juvenile masked boobies—yellow, white and black seabirds with wingspans of five feet—were pulling at the rope with their beaks, loosening the knot. I watched in amusement, then, worn out, fell back asleep.

Back on Tern, my sleep patterns were no better. With no barracks where I could catch up on sleep, the heat and the sound of seabirds often kept me up. The albatrosses clacked and whistled. The boobies whistled and honked. The wedge-tailed shearwaters "wooed" to each other. The sooty terns called in a chorus of deafening tones that sounded like "wide awake, wide awake." As Clemans later puts it when I talk to her about her experience, "Working at night is hard enough when you sleep in a nice, dark, cool, quiet room during the day. But when you have to sleep in a hot, sticky tent surrounded by thousands of squawking birds, that can test your sanity for sure."

During the second week of August, authorities in Honolulu let us know that three tropical storms threatened to turn into hurricanes and come our way. The military helped evacuate biologists from several other remote atolls in Papahānaumokuākea Marine National Monument, but those of us monitoring species in Lalo had the old warehouse on Tern to hole up in, so we stayed. Three of us went to East to take down my camp and had to wade through water packed with floating Portuguese man-of-wars to load the boat. All of us got stung, but one researcher took the brunt of the pain when one of the animals drifted up his board shorts.

The storms missed us, and soon I was back on East, where piles of plastic had washed ashore. One morning, I spied a moving clump of fishing line the size of two fists. A turtle hatchling had crawled into it and was wrapped up. I carefully untangled the youngster and put it down, and it crawled into the surf.

It felt good to help, but I couldn't do right by every animal. One night, a nesting turtle crawled back toward the surf with a large hook stuck in her front flipper. I ran with my multitool to take it out, but the sharp object was driven in deep. I couldn't yank it out before she hit the water.

By the end of the summer, more than 800 turtles had nested

on East, a record for the island. To find them in the dark, I used a small flashlight and all my senses. I looked for moonlight reflected off wet shells. I sniffed the air for the smell of turned-over soil. I felt the sand thrown by mothers on my skin. I listened for the calls of birds that indicated a turtle was bothering its rest. To work them all up, I moved constantly.

But after my time on East was over, I didn't turn off. I couldn't sleep, even after switching to a more normal nighttime routine. A ship took me from Tern to Midway Atoll, then a plane to Honolulu and then back to the mainland, in Denver, to walk my sister Margaret down the aisle. From there, I met with representatives of the EPA in San Francisco—to share what conditions were like on Tern and because I planned to write about their research on microplastics—before returning to Hawaii to finish up my own work. Sleep still eluded me, and my mind began flickering into a manic state. Increasingly, short episodes interrupted my sanity.

It's hard to explain the period that followed. Rather than focusing on finishing up my work and resting, I went on frenzied research tangents. I became obsessed with investigating government testing related to biological weapons in remote areas and wondered if that happened on Lalo. I looked into Smithsonian efforts sponsored by the Army to conduct research on the atoll. I made weird connections between those efforts and my experiences that I wouldn't have accepted in a steady state. Did experiments take place on Lalo? Did weaponized bacteria persist in the animals there?

In that manic state, I flew to New York, where my brain convinced me I would uncover more information that would verify my conspiratorial thoughts. I used a pay phone to make calls and acted like authorities might have been following me. My brother Paul, who was recording an album in the city, thought I was acting weird but didn't know exactly why, and he shrugged it off. One day, on his suggestion, we visited the Metropolitan Museum of Art to see a painting by one of our great-grandfathers.

But when we got to the Met, my mind raced out of control again. I refused to leave a Greek and Roman art gallery. Confused, my brother and his girlfriend at the time left me there. I started examining vases and sculptures in detail. My mind scanned frantically for patterns and signs. Clear as day, the ancient artists' use of animals seemed to show that the creatures were gods, and that now our modern society was taking down those gods. Adrenaline rushed through my body as my thoughts blitzed between objects in the museum, events in my life, and scenes from movies and stories. What was I supposed to do next?

When the museum closed, I ventured outside, my mind still manic. A street performer playing "Gonna Fly Now" from *Rocky* seemed like a sign to run. Clad in jeans, an oversize white T-shirt and brown wingtips, I hit the East River, stopping at a fence and looking at my phone. A man was running toward me; was I being followed?

A text from one of my brothers happened to have the word "jump" in it, and that was enough to make me climb over the fence and leap maybe 30 feet down into the river, still fully clothed with my phone and wallet in my pockets.

As I swam north toward an island called Mill Rock, I waved at people looking out onto the river. I remember noticing that Manhattan was surrounded by a wall, just like Tern. The current swirled in places, and though I felt heavy, I was also on a high. Close to Mill Rock, a boat that I think was helmed by either the police or the Coast Guard approached, and a life preserver splashed next to me. Worn out, but still excited, I grabbed it and was lifted on board. The boat brought me back to Manhattan, where I was loaded into what was likely an ambulance. Workers shrouded me in towels and asked me what I was doing. I said I just wanted to go for a swim.

First responders took me to a hospital, where I was guarded until my brother arrived. He eventually convinced me to sign a form to transfer me to a psychiatric hospital. I weighed about 160 pounds, down from my usual 190. The weight loss had occurred mostly on East, but I also wasn't eating much because of my mania. Health workers kept the door to my room open, and big dudes guarded it through the night. Laces had been taken out of my shoes and drawstrings out of my shorts. Privileges, like wearing my normal clothes, were taken away. I tried to escape by opening windows when I thought no one was looking.

I didn't share much with any of the psychiatrists there, because I still thought I was being pursued, though I hadn't yet determined exactly who was following me or why. Family showed up and brought me cheesecake every day to help me gain my weight back. I ate it, but only after looking for signs about what to do based on the arrangement of mangos and strawberries on top of the dessert.

After about a month of counseling and therapy and drugs, I was released and diagnosed with bipolar disorder. I eventually got a job in California, but I stopped taking my meds and lost my mind again. I got another job, but I went into a deep depression, missed weeks of work, couldn't function properly and got fired. I moved to my mom's basement in Minnesota and regularly slept twenty or more hours a day. My medication made me drowsy and likely slowed my metabolism, but I had to take it if I wanted to recover.

In October of 2018, my state had dramatically improved, and I was working at a small Minnesota conservation magazine when I got an email from Frans Juola. East Island had been destroyed by a storm. "A powerful hurricane wiped out this remote Hawaiian island in a night. It was a critical nesting ground for threatened species," read a headline in the *Washington Post*. Hurricane Walaka directly passed over Lalo as a Category 3 storm and destroyed the island. Only a couple slivers of sand remained above water.

I emailed former co-workers for more info and was mostly left with questions. What would happen to the animals that bred and nested on East? Would more animals move to Tern and get trapped?

Here's what today's researchers tell me about the islands where I once worked long, sleepless nights: East is no longer stable enough for field research over an entire season. And since Walaka wiped out East, more seals and turtles have sought out Tern, placing them in danger of new threats. Take the endangered Hawaiian monk seals, or *'ilio holo i ka uaua* in Hawaiian. With more than 200 seals, Lalo harbors a crucial 20 percent of Papahānaumokuākea Marine National Monument's seal population.

Prior to Hurricane Walaka, almost a third of the monk seal pups were born on East, and another third were born on the nearby island of Trig, which went completely underwater in September 2018, just before East did. With those islands dramatically diminished, mother seals were forced to smaller islands where their pups were more susceptible to drowning in high surf or being killed by sharks. Other mothers moved to Tern. "Monk seals still need to haul out. They still need to have pups. They still need someplace that's safe," says Charles Littnan, the science and research director covering this area of the Pacific for NOAA. "And when all of the other real estate is disappearing, the best place for them is Tern Island—that has a whole lot of danger for them to navigate."

The greatest threat to the seals on Tern is that decaying sea wall. Farry has found seals trapped in holes and cracks of every sort in

the wall, and under mechanical junk likely tossed out by the Coast
Guard. In one stressful situation in 2019, Farry tells me, he and his
colleagues found a seal pup 40 feet into a narrow section of double sea wall that rose about five feet above it. Sand prevented the
pup from going lower. A beam prevented it from going higher—
and the biologists from lifting it out. The tide was rising. "It really
entered my mind that this pup might drown in front of us," Farry
tells me. Biologists spent an hour prodding the animal and shoveling sand out from beneath it so that it could move to a wider
section of sea wall. Once it did, researchers lifted it out.

The animal was lucky. In records that date back to 1989, more
than a third of documented cases of monk seal entrapments have
occurred since 2017. "If islands continue to disappear and seals
continue to shift to Tern Island," says Littnan, "this could be a disaster for seals at French Frigate Shoals [a.k.a. Lalo]."

The situation for turtles is even worse. Scientists think Tern
has become their primary nesting spot due to the reduced size of
the other islets remaining at Lalo, though they don't have hard
numbers because they can't do surveys on East. But in 2019, researchers looked at tagged sea turtles and saw animals that only
ever nested on East Island coming ashore to nest on Tern. During
that season, biologists helped two to three trapped females a week.
"We would do night surveys all night long, and then from sunrise
till eight o'clock we were just trying to move turtles," says Marylou
Staman, a biologist who spent three seasons on Lalo. "That was
incredible. It was exhausting."

In previous seasons on Tern, female turtles often clustered their
nests on the south shore of the island, where bushes and vegetation
proliferated, and soft sand allowed them to dig deep chambers.
With those natural markers wiped away by the 2018 hurricane, the
turtles crawled further inland, where the storm had spread a thin
layer of sand over the runway. They dug through that layer and less
than a foot or so down hit compressed coral as hard as concrete.
Normally the females would dig two feet or more down to lay their
eggs. Now, after hitting rock, many turtles abandoned their shallow
nests and moved somewhere else to dig again. Sometimes they dug
and abandoned such nests for days on end. But not all of the turtles
abandoned their efforts. Some dropped their eggs in the shallow
pits, and the embryos likely cooked from being too close to hot
surface sand.

The competition for space between different species on Tern also ramped up. Before East was hit by Walaka, roughly 4,000 black-footed albatross, 1,000 Laysan albatross and thousands of other birds nested there. Beth Flint, a supervisory wildlife biologist with USFWS who has worked on Lalo since 1980, suspects many of those seabirds crammed onto Tern. She says the increase in turtles and seabirds has likely led to more crushed eggs and chicks. And during the 2019 field season, more than thirty birds got trapped in the sea wall.

Aside from all those entrapments, an invisible danger lurks on Tern: the decaying electrical equipment the military left behind. Before the storm hit, a crew headed by the EPA sampled and monitored Tern for contaminants. They found unacceptably high levels of lead and PCBs. The area with the greatest pollution was the Bulky Dump, that spot where I saw so much debris in 2014.

The Coast Guard contracted a company to do cleanup on the island in the early 2000s, but it apparently did not retrieve everything. PCBs are endocrine disruptors and can be mistakenly accepted by the body as hormones—causing tumors, birth defects and other developmental disorders. USFWS resource contaminants specialist Lee Ann Woodward tells me in an email that almost all of the animals tested on Tern have been contaminated.

So what will be the fate of Tern Island?

Field biologists who have worked there for decades say the island should be returned to a natural state.

My former boss George Balazs, a sea turtle scientist who still actively studies the animals after retiring from NOAA's Pacific Islands Fisheries Science Center, argues it's time to take down the sea wall. "Remove it," he says. "Don't level it and throw in the ocean. You've already thrown enough stuff in the ocean. Let's get rid of it properly with modern equipment."

The Navy wouldn't directly answer my questions about whether it would help pay for the possible removal of the structure. "The Department of the Navy no longer has ownership over Tern Island," it stated in an email. "For questions about the status of Tern Island please contact the Department of Interior/U.S. Fish and Wildlife Service."

I got a similar deflection when I wrote to the Coast Guard, asking if it planned to finish cleaning up the contaminating debris it

left on the island decades ago. "Tern Island is held as property of
the Fish and Wildlife Service who is a federal agency," its lawyers
responded. They tell me that the current owner is responsible for
any cleanup, "even if the contamination was done by another federal agency prior to the current federal agency."

Jim Woolford, former director of the Office of Superfund Re-
mediation at the EPA, disputed this, telling me that the Coast
Guard should in fact be contributing to the cleanup. "Nowhere in
CERCLA [the Comprehensive Environmental Response, Compen-
sation and Liability Act, a.k.a. Superfund] does it say that for federal facilities only the current owner/operator is the responsible
party," he writes in an email. But he adds that no progress would
be made without pressure from the public, environmental groups
or a congressional delegation. "It really has to happen at a very
high level," says Woolford. "And once it gets that attention, things
can move pretty quickly."

The USFWS is still working with other agencies on a plan, according to Jared Underwood, the agency's superintendent for
Papahānaumokuākea Marine National Monument. But the agency
has just about $1 million annually to spend on all lands and waters
in the monument, and it can only allot about 10 percent of that to
Tern. Experts say that cleaning up the contamination on Tern will
run between $2 million and $3 million. Fixing the degrading infrastructure and sea wall will run in the tens of millions of dollars,
if not more. For any sort of meaningful action on that anytime
soon, Underwood says, USFWS is looking for support from other
sources, including possibly appropriations from Congress.

Climate models project that the ocean may rise by two to three
feet, or more, around Tern Island by 2100. And hurricanes like
Walaka may become more powerful and possibly more common
at Lalo as the planet warms. "So the long-term picture for these
islands is bleak, though not entirely hopeless," says Chip Fletcher,
a climate scientist at the University of Hawaii at Manoa who visited
East in 2018.

I asked Todd Bridges, who until this February served as the U.S.
Army's senior research scientist for environmental science and is
now with the University of Georgia, how the island could be protected in the absence of the sea wall. Bridges—who led a U.S. Army
Corps of Engineers initiative called Engineering With Nature—tells
me that a spectrum of interventions could be used to bolster the

island. One solution would be to dredge sand to build the island up. Another might be accomplished in the water, engineering the reef around the island to protect it more from wave action.

Perhaps the geologist with the strongest connections to Lalo is the University of Hawaii's Haunani Kane. The Native Hawaiian is a navigator who first visited the atoll as a twenty-year-old, arriving on a voyaging canoe guided only by the stars and other environmental cues. She returned to Lalo more than a decade later, in the summer of 2018 with Fletcher, to study East Island and its relationship to sea-level rise.

Before dredging sand and modifying reefs, Kane thinks scientists need to understand more about the natural relationship between islands and reefs. "The last thing you want to do is manipulate and engineer the system in a way that takes away its natural resiliency as well," Kane says. In 2021, she and her team saw the island come back to about 60 percent of the size it had been before Hurricane Walaka nearly wiped it off the map in 2018, though it's less stable and a shadow of its former self. Reefs grow and then degrade to become sand and chunks of rock that build islands up.

For Native Hawaiians, Lalo is a cultural resource as well as a natural one. "It's a place where we believe our *kupuna*, or our ancestors, go as we transition into the next realm," Kane says. "We view these islands as not just a place, but as a place of our ancestors, and we view the islands as our ancestors themselves."

Pelika Andrade, a Native Hawaiian intertidal, watershed and cultural researcher who sits on the reserve advisory council for the monument, tells me that the atoll was a bucket list run for a generation of fishers, including her grandfather. She sees what has happened to Tern as indicative of the problematic nature of colonialism. "There's a reason why there's so much distress in the system, because historically this is the repeat, right?" she says. "Set up something, but the plan of taking it down? There's no long-term plan. And then there's kind of like an abandonment that happens, and others are tasked to take care of the mess."

Researcher Kevin O'Brien is one of the people trying to take care of that mess. He arrived on Tern in October 2020, after a season when the island had been empty of biologists due to the Covid-19 pandemic. O'Brien had formed a nonprofit called the Papahānaumokuākea Marine Debris Project to collect ocean trash, but he traveled to Tern to address degrading infrastructure scattered across the island by Hurricane Walaka. He brought along a crew that included a welder, a metalworker, some former construction workers, heavy equipment drivers and a handful of biologists. He also brought a skid steer, a utility vehicle, a trailer, jackhammers, generators and several metal-cutting tools. They found numerous dead seabirds stuck in the sea wall and dozens of hatchlings that the sun had dried into jerky. During that trip and one earlier in the year, workers found seven trapped dead adult turtles.

O'Brien's team cut eight large holes in the sea wall so seals and turtles could escape. They jackhammered concrete to break up gaps where sea turtles could get stuck, and they built a fence to prevent turtles from crawling to an area of the island where the animals could get trapped. They cut up lumber, cables, fiberglass, roofing, a 20-foot shipping container, three derelict boat trailers and other material that was thrown across the island. After ten days, they carted 82,600 pounds of garbage off Tern, clearing hurricane and marine debris from almost 22 acres of land. Though USFWS and NOAA served as partners, O'Brien's crew did what government agencies had been unable to organize and complete alone.

But the nonprofit left with a lot unaddressed. The barracks, warehouse and generator shed—which had leaking batteries and fossil fuels—were falling apart. More than 100 pieces of large black pipe meant to serve as barriers to crawling turtles were scattered about the island. All of those things could injure or kill more wildlife, especially if another storm comes through.

O'Brien knows that his efforts were just stopgap measures. Too many large and evolving hazards exist. In 2021, even after his helpful work, juvenile green sea turtles became trapped nine times, and nesting turtles were trapped at least fifty times. At least seven of those adult females died. Seabirds and critically endangered Hawaiian monk seals were also trapped by hazards on the island.

In 2022, the rate of entrapments was just as bad.

Considering the toll this work takes on researchers, and the overwhelming forces at work, why does Tern Island matter? After all, the place is just a pinprick on a wall-sized map of the world.

It matters because this tiny speck in the ocean has a far-reaching influence. Many of the animals birthed there bring benefits to habitats far away. Take the sea turtles, who migrate hundreds of miles to the main Hawaiian Islands, where they feed on algae and keep coastal ecosystems in check. All over Waikiki, signs and brochures advertise expeditions to snorkel with the marine giants. Souvenir shops feature sea turtles on cups, flip-flops, magnets and more. The animals, known as *honu* in Hawaiian, also factor prominently into Native culture. To determine how to best protect this important species, authorities take into account what scientists find out about the nesting population size on Lalo. And seabirds on Tern roam even farther, providing important services as far as the waters off California and Alaska, including fertilizing the land and ocean with their guano, thus spurring the growth of plants, coral reefs and phytoplankton at the bottom of our food web.

Of course, I am personally invested in Tern's long-term viability. While there, I interacted daily with resilient but vulnerable animals struggling against human threats that added a degree of difficulty to their survival. And perhaps I see connections between Tern's damaged state and my own once damaged state.

During my darkest time in the psychiatric hospital, I was lost, out of my mind and illogically scared. I was surrounded by others in similar states—and worse ones. Patients babbled incoherently, melted down while imagining unseen threats and gazed off despondently in the lowest conditions of their lives. In the middle of all of this, overburdened health care workers tended to them.

One of those health care workers stood out—a small Jewish woman in her sixties named Karen Wald Cohen. She was energetic and engaging, and often wore outrageously colorful outfits—oranges and pinks and yellows and rainbow-themed get-ups—that burst amid the bland scrubs and socks many patients wore. She went up to depressed patients sitting alone and shared personal stories. She quietly reasoned with and talked down grown men half her age and twice her size during angry outbursts.

In my worn-down state, I thought there was something weird about her, but I liked it. I didn't realize this then, because everyone inside kept telling me to focus on myself, but I now realize that her actions were some of the bravest exploits I'd ever witnessed. In some ways, her efforts to care for severely disturbed people weren't all that different from the efforts of researchers on Tern trying to free thrashing animals. She worked as the last line of defense against much larger societal problems.

Karen took care of all of us, even people whose own families had written them off as lost causes. Aside from my visits with my family, my conversations with her, where she shared sketchy stories from her life and hilarious episodes capped by her boisterous laugh, were the most healing parts of my stay.

One day, Karen strolled down the tiled hallway wearing an almost completely black outfit. I think the bottom of the dress was frilly and layered, like a tutu, and she had on large black leather boots. Little shiny dots were scattered about her outfit, but in my confused state, the blackness threw me. It was not just that it was a weird get-up for one to wear into a psych ward, but it also didn't fit with her normally colorful outfits and personality.

Months later, when recovering in Minnesota, I got an unexpected letter in the mail from Karen. It was something she didn't have to send; I was no longer in her care. I opened the envelope and pulled out a card, pitch-black on the outside, with scattered shiny dots.

The card reminded me of the nights I'd spent on the islands of Lalo. In the deep blackness of the sky, unmarred by light pollution, stars glimmered from horizon to horizon. Rarely seen celestial phenomena stood out. One night on East, I rounded the northwest corner of the island and saw an arch extending up from the ocean. A moonbow—as big as any rainbow, with white gradations instead of colors—interrupted the darkness.

Humans rarely get to experience nights like the ones I had on Tern and East Islands, but any researcher who has spent time there will tell you that our species is worse off without those experiences of awe. We're connected to all the teeming species that venture out from those distant islands, and their struggles with plastics, ruins and disappearing land are more and more becoming our own.

It can be difficult to work in a remote environment where such threats are so stark, to fight them and think about them on a daily basis, without succumbing to exhaustion or even madness. But scientists keep going back, year after year, because they believe it's worth it. Like me, they hope that telling the world about the devastation happening in the middle of the Pacific Ocean will help people everywhere realize what we can save.

Or, as the message inside of that card from Karen read:

It takes darkness to see the stars.

EMMA MARRIS

The Sea Eagles That Returned to Mull

FROM *Hakai*

SHE COMES WINGING in from behind us, looming into our field of vision, seeming almost too massive to be airborne. She is a white-tailed eagle, one of a species of sea eagles. *Haliaeetus al bicilla* is a close cousin of the North American bald eagle, with its same dour expression, outsized muppety beak, and slightly ramshackle habit of motion, landing like a winter coat falling off a hook. The wing-span of a big female can reach 2.5 meters. These are mythically big animals. Their size makes them bold. They lack the furtive elegance of so many other wild animals. They look casual, like they own the place.

The place owned by this particular eagle is the Isle of Mull, a rugged island off the west coast of Scotland. I am sitting in a truck, parked near a small copse of spruce. Next to me, with a scope mounted on his windowsill, is Dave Sexton from the Royal Society for the Protection of Birds, who has studied and protected eagles on and off on Mull since the 1980s. As we watch the eagle approach her nest, we see that she carries a twig in her yellow di-nosaur talons. Sexton explains that this pair lost their chick a few weeks ago when a spell of cold and wet weather happened to hit the island just after hatching. With their nestling dead, the couple seem lost. Although white-tailed eagles almost never lay a second clutch, the pair add sticks to their already built nest, perhaps com-pelled to do so by the stimulus of it being empty.

This pair of eagles have raised several chicks in years past. A local sheep farmer named Jamie Maclean had complained that they were raising their chicks on a steady diet of his newborn lambs, which are born in spring, just as chicks hatch. And so, with Sexton's help, the Scottish government agreed to pay for some "diversionary feeding." Maclean would buy fish from a local fishmonger—at retail prices—with government money, and then put them out for the eagles to eat. The idea was that with the free fish rolling in, they'd leave the lambs alone.

"So you can't quite see it," Sexton says, pointing out the window of the truck. "But just on the other side of that trailer, up on the hill, there's a very prominent mound in the middle of a field." It is here that the farmer offers up his fish.

"You realize this sounds like some sort of Iron Age ritual?" I ask.

"It does a bit," he says, chuckling.

Sexton believes that eagles take far fewer lambs than other birds, such as hooded crows, ravens, and black-backed gulls. But he still thinks the diversionary feeding program is worth the expense— even if it is mostly a gesture of goodwill. "If I can do something like help with the diversionary feeding . . . the fact that the eagles are taking fish and not hanging around the lambs, that's a good thing."

I am planning to meet with Maclean. I am curious to know how he feels about this pair, after feeding them for years and watching them raise several generations of chicks. I wonder if their intimacy has made him feel any warmth toward them, despite their predatory interest in his lambs.

White-tailed eagles once lived across most of Eurasia. As far as scholars can tell by the hoards of their talons in at least one burial site, in antiquity they were symbolically potent, but as Europe became increasingly agrarian, they gained a reputation as a threat to livestock, and they were shot and poisoned out of most countries in western Europe in the nineteenth and early twentieth centuries. The last well-attested eagle seen in Scotland was shot in Shetland in 1918.

Then, starting in 1975, conservationists reintroduced sea eagles from Norway to the Scottish Isle of Rhum, a nature reserve about 30 kilometers north of Mull. By 1985, a single pair had arrived in Mull and constructed a large sprawling nest in which a chick successfully hatched. Today, there are some twenty-two mating pairs on Mull, which is about 875 square kilometers, roughly the size of New York City.

When the eagles first returned, sheep farmers were not happy. Their grandfathers had killed these birds for a reason. But now nearly forty years have passed—more than a generation—since the birds returned. I went to Mull because I wanted to see whether time had softened the acrimony that can accompany a reintroduction of a controversial animal. Would younger farmers who had never known a Mull without eagles still see them as an unwelcome presence, imposed upon them by clueless citified nature worshipers? Or would they have gotten used to them, enjoyed the fruits of the tourism they attract—up to £8 million (US $10 million) according to one estimate—and maybe even learned to love them a little bit?

The next day, I drive to meet Jamie Maclean at one of the pastures where he keeps his sheep. Mull is a strange landscape, both rugged and manicured. On the one hand, it feels quite wild and remote. Sheer cliffs plunge into the sea; oystercatchers and lapwings strut around the empty beaches, and in the steepest ravines, where sheep fear to tread, there are fairy-tale woodlands, with mossy oaks, dripping ferns, and carpets of bluebells and celandine in the spring. But much of Mull is used as grazing land for sheep and cattle, and they do a thorough job, keeping the hills nibbled short. It looks a little bit like the perfectly round grass hills on the set of the iconic children's show *Teletubbies*, except inhabited by sheep instead of slightly eerie anthropomorphic puppet people.

Maclean hops out of a Toyota Hilux. He is thirty-two and fresh faced, and we go for a stroll through one of these lawn-like landscapes. As we move slowly through his sheep, they gently adjust their position, keeping calm but also keeping a healthy distance from us. Maclean tells me that there are ten breeding pairs of eagles on the ground he and his uncle pasture their sheep on. At this site, where the flock comprises four hundred breeding ewes, he figures they lose maybe thirty lambs a year to the eagles. "It is so hard to tell," he says. In the style of farming employed on Mull, known as hill farming, ewes are often scanned by ultrasound in the winter to see if they are carrying one or two lambs. They typically give birth on their own, in the hills. Later, the farmers count all the singles and twins and compare the count with the predicted number based on the scans. But some lambs are born dead or die of disease or cold. Crows and jackdaws and gulls peck at sickly lambs. Some fall off cliffs. Determining which ones disappear due to eagle predation is difficult. The proportion of lambs that simply disappear is known as *black loss* and the fraction of that that is due to eagles remains a source of endless debate.

Since 2016, Maclean has put out fish for some of the eagles and watched them swoop in to dine when he was just 15 or 20 meters away. "They aren't really scared," he says. But he reckons the feeding did reduce his lamb losses.

I ask him how he felt about the pair I'd seen when this year's chick died. Did he feel sad for them? He smiles at the thought. He says he is "not too fussed really." "Mull is at capacity for eagles," he adds. It doesn't need more. Sexton would quibble with him, arguing that a few more pairs could probably fit on the island. But there are indubitably a lot of eagles on Mull. I ask Maclean, If you were presented with a magic button that would get rid of all Mull's eagles instantly, would you press it? He thinks about it for some time, then answers that he would not. Some of his friends benefit from birding tourism. "I wouldn't do anything to get rid of them. They are pretty incredible birds." But he wouldn't mind having the right to get rid of individual eagles that are particularly troublesome. "Some birds are a lot worse than others."

I hear very similar sentiments from all the sheep farmers I speak to on my visit. Eagles are here to stay on Mull and provide economic benefits in the form of tourism revenue—often to the sheep farmers themselves, most of whom are also renting out vacation cottages. But the farmers still don't really like them. And how could they? The central purpose of a shepherd's life is to raise sheep. Yes, the sheep are destined to be killed and eaten, but until they are slaughtered, they are in the care of the farmer, who must keep them alive and healthy at all costs, around the clock, in the wet and cold. Seeing a massive raptor flying over the field with the bloody carcass of one of the tiny fluffy lambs you just spent a year trying to nurture and protect feels like a punch in the gut. Proposed fixes are, generally speaking, dismissed by the farmers I speak to. Bringing the entire flock in to lamb indoors is impractical for many operations and spreads disease. Additional shepherds are hard to find and too expensive. And so on.

Farmers are particularly annoyed that conservationists long denied that eagles ever preyed on live lambs. Sexton and other bird experts acknowledge that this may occasionally happen, but they still maintain that the eagles are much more likely to scavenge a lamb that has died of something else or steal a lamb from another predator than to attack a perfectly healthy lamb. Sexton's line is that they are seldom the ultimate cause of death. "Whilst they might have been seen carrying the lamb off, or indeed feeding on it, and it might even still be warm, they didn't start the process, right? They are taking advantage of something that other predators have started," Sexton says, pointing out that gulls, ravens, and crows are well-known lamb killers. Farmers tend to disagree. Many of my interviews quickly devolve into discussions of exactly how many lambs eagles kill, and I can hear people—farmers and eagle advocates alike—slip into bitter grooves, reciting arguments to me that they have made many, many times before. The eagles, it appeared, had not become an unremarkable part of the flora and fauna over time. They are still making everyone mad.

I sit on a bench overlooking the harbor in the almost comically picturesque town of Tobermory and watch Italians and Londoners with binoculars slung around their necks eat fish and chips and snap selfies. As an environmental journalist who has written frequently about wolf reintroduction in North America, I am a bit depressed. I had been so hopeful that after all these years the eagle would have been thoroughly reintegrated into both the ecosystem and the human-nonhuman community of Mull. In my naive optimism, I assumed that as soon as a generation of farmers grew up in a world with eagles, they would take them for granted as a rightful part of the island's life. But instead, they nurse a grudge, against the birds and the birds' advocates. Because every time a glimpse of a magnificently soaring eagle makes them decide that maybe the birds aren't so bad, they come around a corner on their quad bikes and find a pile of guts and wool and wonder if an eagle has just blithely killed yet another of their lambs.

I think back to a conversation I had had on the train toward Mull. A young woman told me that the first time she had taken the journey through the Highlands, she had become overcome by the beauty of the landscape and had felt an almost out-of-body sensation, as if she were "flying alongside the train." I was happy for her, but when I looked out of the window, I saw a landscape that looked a lot like my home in Oregon—except denuded. The basics looked right: the steep mountains, the little rocky coastal islands, the moist air. But instead of being clothed up to their summits by a tangle of temperate rainforest, the mountains were covered with sheep lawn. There are extensive timber plantations in the Highlands and on the Hebrides as well, including on Mull, but the endless rows of same-aged Sitka spruce don't host much biodiversity and don't look quite right to my eye either.

On my journey to Mull, I picked up a copy of *The Lost Rainforests of Britain* by Guy Shrubsole. He argues that some 20 percent of Britain—much of the western coast, including especially the Scottish Highlands and Hebrides—was once temperate rainforest, similar to the old-growth forests in Washington and British Columbia. According to the map in the book, thousands of years ago, Mull could have been nearly all forest, a woodland of pine, oak, hazel, and elm. And presumably any white-tailed eagles living there at that time would have lived a life similar to forest-dwelling bald eagles in the Pacific Northwest: nesting on tall snags above the canopy, working the forest-beach edge, fishing for salmon and mackerel, hunting ducks and seabirds, stealing prey from other predatory species, maybe nabbing the odd otter pup. Today, while fish and birds are available to white-tailed eagles, they are also confronted with a vast swath of open land, populated, each spring, with newborn lambs taking their first wobbly steps, exposed and knock-kneed, weighing just 2.5 kilograms. It would be kind of bizarre if they didn't eat some lambs.

Returning to a world before sheep is not the goal of most British conservationists—nor should it be. Nature is dynamic, not static, and it is neither possible nor desirable to simply roll back all human changes on the land. But taking a deeper time perspective helps me see that there have been many Mulls over the millennia, and it enlarges my sense of what might be possible in the future. I realize I had been expecting some kind of frictionless coexistence to develop on Mull purely based on goodwill. Maybe hill farming was simply not fully compatible with dense eagle populations. But then which use of the land should have priority? Sheep or eagles? White-tailed eagles are not globally endangered. They are so common in Russia, Norway, and China that the species is ranked as least concern by the International Union for Conservation of Nature. So the argument for their presence isn't that the white-tailed eagle needs to be *saved*. The argument is these sea eagles *belong* on Mull. But many would also say that sheep belong on Mull.

Like many countries, Scotland subsidizes its farming industry to ensure some level of domestic food security, to prop up rural economies, and to protect cultural landscapes. The Scottish government heavily subsidizes hill farming, paying farmers £61.25 ($77.06) a head for ewes. Although there are policies in place to avoid extreme overcrowding, the per-head payments encourage farmers to run many animals, creating intense levels of grazing. Where grazing is heavy, soil can be eroded and compacted, plants struggle to flower and set seed, birds who feed on insects have nothing to eat. And yet lamb prices are so low that without subsidies, the whole enterprise would be unprofitable. The wool is worth so little it is often simply thrown away or burnt. Hill farming is even less profitable than lowland sheep farming. One of the routes by which the Scottish government subsidizes hill farming is literally called the Less Favoured Area Support Scheme, an admission that these areas are simply less suited for producing lamb.

Some Scottish farmers are also enrolled in agri-environment schemes, which pay them to manage their land for biodiversity. Maintaining traditional diverse hedgerows, delaying grazing fields until ground-nesting birds like corncrakes have fledged, building bridges for livestock so they won't trample stream banks—all these and many more actions can earn money for farmers. But with the per-head ewe payment calling, the end result is that taxpayers fund opposing government actions: a largely unprofitable agricultural industry that tends to destroy biodiversity and relatively small-scale efforts to mitigate that destruction.

In large part, the continued support for hill farming is cultural. People expect the Scottish countryside to be close shorn and sheep dotted. In 2009, a representative of the National Farmers Union warned of the consequences if sheep farming were to continue to decline on Mull, telling a government minister, "As an island, we are also very reliant on our tourist trade but our beautiful scenery is also at risk if abandoned land is allowed to run to scrub and bracken rather than being grazed and managed."

Some tourists might actually appreciate the scrub, especially those being guided by Ewan Miles. Miles has spent thirteen years on Mull, working in nature tourism. He began working on a tourist boat, then opened his own land-based wildlife watching business in 2015, beginning with a single seven-seater vehicle. Today, he has two nine-seater vehicles and two seasonal naturalists running tours. Miles has thought a lot about the dominance of hill farming. And he imagines a different way. What if instead of sheep, farmers raised pigs and cattle in densities that would echo the way the wild boar and the extinct cattle ancestor known as the auroch may have once acted on the landscape? And what if revenue from these animals was just part of a property's income, along with payments for carbon sequestration and biodiversity? Baby Highland calves and piglets would be less vulnerable to eagles, and with less grazing there would be more habitat for hares and grouse, which could be alternative prey for eagles.

I am thrilled by his vision. A solution! A way forward. Maybe sheep are, after all, *not* the future of Mull—although I can well imagine this idea would not be enthusiastically welcomed by sheep farmers themselves. I ask Miles who are the most open-minded and biodiversity-oriented sheep farmers he knows. I want to speak to someone who might represent the future instead of the present, which seems to be dominated by an endless loop of arguments about eagles and lambs. He suggests a couple, Carolyne and Somerset Charrington at Treshnish Farm. They raise lambs but also host guests in beautiful rental cottages—guests who are often traveling to Mull in part to see eagles.

On the road to Treshnish, I drive past a well-known eagle watching hotspot called Loch Na Keal. All along the shore of the lake, people with scopes and binoculars sit and wait, scanning the skies for those big, rectangular outspread wings, dark and squared-off, like a flat-screen TV. Some are in lawn chairs outside their camper vans; others are perched on rocks like big birds themselves. There is something reverent about them, as if they are pilgrims awaiting a visitation.

I am expecting the Charringtons to be eagle fans, given the way Miles described their commitment to caring for nature, but I am immediately disabused of this expectation. Even before our cups of tea are ready, they have begun explaining how eagles are a much worse menace than the conservationists make out. Somerset explains that when lambs simply vanish, eagles are the most obvious explanation. After all, he says, "the only way out of the field is up." The couple are also a bit skeptical of my wild-eyed ideas about reforesting Mull, creating wet and mossy rainforests where eagles could thrive. Stop grazing and all you would reap would be an endless monoculture of bracken fern, they say. Getting an entire diverse forest back would take forever.

Frustrated, I push the Charringtons, demanding some kind of solution to the conflict. "There is no solution," Somerset says. "The eagles are here, and they're here to stay. Full stop. And we have to farm within that."

I find this response hard to process. *No solution?* Farmers will just continue to raise sheep, eagles will continue to eat them, farmers will remain furious and angry, and it will just go on like that forever?

As we walk out of the house and toward the lambing barn, where ewes giving birth for the first time are housed so they can be helped, Carolyne says that each lamb lost to an eagle hurts her deeply. And yet, like so many other farmers, she can also see their charm. "I hate what they do, but they are still jaw dropping when you see them flying," she says. Then she tells me that the way she and Somerset farm is heavily shaped by government farm policy. If government farming policy shifted to favor cattle over sheep, or to put more money in biodiversity than in ewes, they'd shift along with it.

The Scottish government is in fact in the midst of a process to revise its agricultural policy now that the United Kingdom is no longer part of the European Union. Mairi Gougeon, cabinet secretary for rural affairs, has promised a "robust and coherent framework to underpin Scotland's future agriculture support regime from 2025 onwards, that delivers high quality food production, climate mitigation and adaptation, and nature restoration."

The next day, I follow a woodland trail from Tobermory to a lighthouse on a point. Clambering around on a rocky beach, I find a dead gull, its bloody head popped off its shoulders but still attached by a snaky length of exposed vertebrae. I smell the sweet smell of death. Was it a sea eagle at work?

With the stench still in my nose, I consider that many countries have recently decided to reintroduce predators, inviting them back to lived-in landscapes despite knowing full well how they make their living. Is this a mark of ecological maturity? Or is it perhaps also a bit of showing off—a bit of "conspicuous conservation" by which we show off the affluence of our societies. "See?" we say. "We are so well off that we can let these wild animals feast on our livestock and absorb the loss."

There is an inherent violence in biodiversity. Behind that dry-sounding term are complex food webs: plants eating sunlight, herbivores eating plants, predators eating herbivores, scavengers and detritivores eating the leftovers. Death—often violent death—is the mechanism by which energy flows from individual to individual, from species to species. Death is also the chisel that carves out new biodiversity, the mechanism by which natural selection slices away "unfit" variants to reveal new genetic combinations: new species.

Somerset Charrington was correct in a very deep sense when he said there was "no solution" to the conflict between eagles and lambs. The Scottish government could retool its agricultural and environmental policies to reduce the dominance of sheep farming in many areas. It could spend more to pay farmers to produce and maintain biodiversity and sequester carbon. The Mull of the far future might have significantly fewer sheep, perhaps more cattle or pigs, more rainforest. But if eagles are to stay, it will not have less death. Whether it be a lamb, a hare, a grouse, or a mackerel, the eagles must eat.

How Do I Make Sense of My Mother's Decision to Die?

FROM *The Atlantic*

MY MOM COULD always leap into the coldest water. Every summer when we visited my grandma in upstate New York, my mom dove straight into the freezing lake, even when the temperature outdoors hit the 50s. The dogs, who usually trailed her everywhere, would whine in protest before paddling after her, and the iciness left her breathless when she surfaced. "Just jump, Lil," she'd yell to me, laughing, before swimming off to vanish into the distance.

But I never could. I didn't think much about that difference between us, until I flew north to be with her on the day she'd chosen to die.

When my mom found out in May last year that she had pancreatic cancer, the surgeon and the oncologist explained to our family that cutting out her tumor might extend her prognosis by about a year; chemotherapy could tack on another six months. A few days later, my mom asked if we could spend time together in Seattle over the summer, if we could get lemonade at the coffee shop while I was there, if I wanted to play Scrabble before I left. "Yeah, of course," I said. "But—" She interrupted me: "I'm not getting surgery."

After a decade of Parkinson's disease, my mom already experienced frequent periods of uncontrolled writhing and many hours spent nearly paralyzed in bed. That illness wounded her the way losing vision might pain a photographer: Throughout her life, she had reveled in physicality, working as a park caretaker, ship builder, and costume designer. Now, plagued by a neurological disorder that would only worsen, she didn't want to also endure postoperative wounds, vomiting from chemo, and the gloved hands of strangers hefting her onto a bedpan after surgery. Nor did she want to wait for the pain cancer could inflict. Instead, my mom said, she planned to request a prescription under Washington's Death with Dignity Act, which allows doctors, physician assistants, and nurse practitioners to provide lethal drugs for self-administration to competent adult residents with six months or less to live.

As a doctor myself, I've confronted plenty of death, yet I still found myself at a loss over how to react to my mom's choice. I know that the American tropes of illness—"battling to the end," "hoping for a miracle," being "a fighter"—often do harm. In clinical training, none of us wanted to unleash the fury of modern medicine upon a ninety-eight-year-old with cancer who'd just lost his pulse, but we all inflicted some version of it: ramming his purpled breastbone against his stilled heart, sending electricity jagging through his chest, and breaching his throat, blood vessels, and penis with tubes, only to watch him die days later. I didn't want that for my mom; I had no desire for her to cling futilely to life.

And yet, even though it shamed me, I couldn't deny feeling unnerved by my mom's choice. I understood why she'd made it, but I still ruminated over alternate scenarios in which she gave chemo a shot or tried out home hospice. Though her certainty was comforting, I was also devastated about losing her, and uneasy about how soon after a new diagnosis she might die.

My mom had made her end-of-life wishes known by the time I was in fifth grade. Our rental home still held the owners' books, among them *Final Exit*, a 1991 guide for dying people to end their lives. The author dispensed step-by-step advice on how to carry out your own death, at a time when nothing like the Death with Dignity Act existed in any state. When I found the book, my mom snatched it away. But months later, after her best friend died of brain cancer, she asked if I remembered it.

"If I ever get really sick, Lil," she said, "I don't plan to suffer for a long time just to die in the end anyhow. I would take my life before it gets to that point, like in that book. Just so you know."

After her Parkinson's diagnosis, my mom moved across the country to Washington, mostly to be near my sister, but also because in 2008, it became only the second state to approve lethal prescriptions for the terminally ill. Since then, despite much contention, the District of Columbia and eight more states have followed—including California, where I live and practice medicine. No dying patient of mine had ever requested the drugs, so I didn't think much about the laws. Then my mom got cancer, and suddenly, the controversies ceased to be abstract.

Proponents of aid-in-dying laws tend to say that helping very sick patients die when they want to is compassionate and justified, because people of sound mind should be free to decide when their illnesses have become unbearable. Access to lethal medications (which many recipients never end up using) lets them concentrate on their remaining life. I sympathize: I've seen patients who, despite palliative care, suffered irremediable existential or physical pain that they could escape only with sedating doses of narcotics.

But I grasped the other side of the argument as well: that self-determination has limits. Aid-in-dying opponents have said that doctors who hasten death violate the Hippocratic Oath. Although I disagree with these moral objections, I do share some of the antagonists' policy concerns. Many worry that state laws will expand to encompass children and the mentally ill, as they do in countries such as Belgium and the Netherlands. They argue that a nation that still devalues disabled people needs to invest in care, rather than permit death and open up the risk of coercion. So far, Americans who have used these laws have been overwhelmingly white and college-educated. But I could imagine patients of mine requesting death for suffering that's been amplified by their poverty or lack of insurance.

These policies are so polarizing that people can't even agree on language. Detractors refer to "assisted suicide," or even murder, while supporters prefer medical "aid-in-dying," which I'll use, because it's less charged. But I don't much like either term, and neither did my mom. She was already dying, so she didn't think of her death as suicide. Nor would she accept a passive term such as "aid-in-dying," when she was the one taking action. Lacking any suitable word, she settled on a phrase that felt stark but honest. "When I kill myself," she'd say. When she killed herself, we should give her spice rack to a friend. When she killed herself, we shouldn't hold a funeral, because that would be depressing. Her tone was always matter-of-fact. My stomach always somersaulted.

That summer, I read constantly about aid-in-dying—accounts of its use in Switzerland, essays in American medical journals, articles written by people who'd lost a loved one that way. I was the exception in our family. My mom was concerned with bigger issues, like whether the ice-cream shop would restock the lemon flavor before she died. My sister thought I was overintellectualizing things—and she was right. Sometimes we do the only thing we know how to, to keep from falling apart.

So I kept looking for the solace of stories that felt as complicated as my own thoughts. They were remarkably rare. To me, loving my mom meant acknowledging my own hesitation yet still respecting her measure of the unendurable. Juggling these emotions felt nuanced, but most of what I read didn't. So many narratives cast aid-in-dying as either an abomination or the epitome of virtue, in which a dying person could be rewarded for courageous serenity with a perfect death.

Another daughter whose mother pursued aid-in-dying spoke in a TED Talk of the "design challenge" to "rebrand" death as "honest, noble, and brave." But however tantalizing the prospect, the promise that we can scrub death of ugliness felt dangerously dishonest. Death can be wrenching and awful no matter where and how it happens: on a ventilator in an intensive-care unit, on morphine in hospice, or with a lethal prescription at home, surrounded by family. Being able to control death doesn't mean we can perfect it.

The myth of the "good death"—graceful and unsullied, beatific even—has infiltrated the human subconscious since at least the fifteenth century, when the *Ars Moriendi*, Christian treatises on the art of dying, proliferated in Europe. A translation of one version counsels the sick on how to die "gladly." The moral in these texts bludgeons you: How you die is a referendum on how you lived, with only a picturesque exit guaranteeing repose for the soul.

The notion has seeped through generations. "I hope if I'm ever in that situation, I'd have the bravery to do that," one friend said about my mom's choice. "It's good she'll die with her dignity intact," said another. My mom's physicians, kind and smart people, seemed so eager to validate her decision that the aid-in-dying criteria distilled to a checklist rather than unfurling into conversation. Even the name of the law my mom intended to use, Death with Dignity, implies that planned death succeeds where other ways of dying don't. More than half a millennium after the *Ars Moriendi*, we still seem to believe that you can fail at death itself.

One doctor told us of a landscape architect who drank the fatal cocktail while exulting in her garden in full bloom. It sounded perfect—except that in all my years as a doctor, I've never seen a perfect death. Every time, there's some flaw: physical discomfort, conversations left unfinished, terror, family conflict, a loved one who didn't get there in time. Still, my sister and I tried to stage-manage a beautiful death. We booked a cabin in Olympic National Park for my mom's exit. We would bake her famous olive bread and cook bouillabaisse. We'd wheel her to the beach, then to the towering cedar forest, then massage her feet with almond oil while we talked in front of a woodstove. The fireside conversation would be our parting exchange of gifts, full of meaning, remembrance, and closure.

As our family waited for that day to come, we kept thinking we should be tearing through a bucket list. Instead, we did what we always had—cooked, played games, read. We just did it with an ever-present sense of countdown, in an apartment where nearly everything would outlive my mom: the succulent on the windowsill, the lasagna in the freezer she made us promise to eat when she was gone.

My mom did have the lemon ice cream again, but our family never made it to the cabin in the forest. A month before the planned trip—ten weeks after my mom's diagnosis—the pharmacy compounded the drugs: a mixture of morphine and three others. The bottle was amber, filled with dissolvable powder and labeled with the words NO REFILLS. ("Now that would be a dark *Saturday Night Live* skit," my mom told me.) The next morning, a Thursday, she called, dizzy and miserable. She wanted to die ahead of schedule, on Saturday. I got on a plane.

My mom, my sister's family, and I spent Friday grilling chicken and drinking good wine. After my older niece painted my mom's nails lavender with polka dots, the kids and my brother-in-law said their goodbyes and left. The next morning, my sister and I laid

out the backyard like a set: a couch swathed in blankets. Tables
with plants and photos and huge candlesticks. A stereo to play the
music of our childhood and her motherhood.

But our revised choreography couldn't erase how horrible my
mom felt that morning, dispirited by her disease and deeply exhausted. We had to cajole her not to die in bed. Eventually, she
came outside, where we drank peppermint tea and talked about
nothing memorable. When the moment came to gulp the bottle's
contents, mixed into lemonade, she didn't hesitate.

"You would make the same choice if you were me, right?" she
said, setting down the empty bottle. I knew she wasn't second-
guessing. She was ending her time as our mother not out of lack
of devotion, but because all other options felt untenable, and she
needed confirmation that we knew this.

"Yes," my sister said, "I would."

"Me too," I said—but in truth, I didn't know. Maybe I would
have dwindled over months of chemo as I learned to reshape my
life in the face of imminent death. Maybe I would have died in
hospice, surrendering myself to the fog and mercy of morphine.
Maybe I would have stowed the drugs in a cupboard, cradling them
occasionally and then, unable to reconcile the simplicity and complexity of that ending, replacing them. Each of these paths would
have demanded its own form of courage—just not my mom's type.

"I'll just go to sleep now, right?" she asked.

"Yeah, Mom, you'll just go to sleep," I said. "I love you."

My sister and I kissed her forehead, her cheeks, her collarbone.
We avoided the poisonous sheen on her lips, where our tears had
wet the residue of white powder.

The aspens rustled, confetti of silver. My mom didn't cry, and
the slightest trace of a smile alighted on her face.

"Bye," she said. "You've been awesome."

And then she dove off the dock. Her lips blued, and when she
tried to speak more, the words never surfaced.

It took her five and a half hours to disappear completely, while
my sister and I tamped down growing worries that the drugs
hadn't worked. My mom felt no pain—she couldn't have, after all
that morphine—but her passing wasn't a fairy tale. Her suffering
wasn't embossed in meaning; she didn't tile over her bitterness
with saintly forbearance. My mom died on the day she was ready
and by the means she chose. All of that matters, immensely so.

She also died precipitously, far from the forest she'd dreamed of, while my sister and I were left with little closure and a prolonged, confusing death.

Usually, I write when I'm most upset, but my mom's death catapulted me into a frightening depth of wordlessness. Weeks passed before I realized that my problem was not that I couldn't find words at all. It was that I couldn't tell the tale I felt I was supposed to. In that myth, death has a metric of success, and that metric is beauty. The trouble is that you can't grieve over a version of events that never happened. You can only grieve over the story you lived, with all of its ambiguities.

My mom's death was beautiful. It was also terrible, and fraught. That is to say, it was human.

RACHEL MAY

The White Oak Tree at McLean:
A Case of Recovery

FROM *Arnoldia*

> Those who have learned to listen to trees no longer want to be a
> tree. They do not yearn to be anything but what they are. That is
> home. And that is happiness.
> —Hermann Hesse

DRIVING UP THE hill to enter campus, the road gently arcing to
the left, it's always the trees that woo me: first, an apple orchard to
the left, then, beside the first of the brick Gothic and Tudor buildings that lend this place its stateliness (or ghostliness, depending
on who you ask), a row of massive pines standing sentinel, shading
the small groups of patients who sometimes take advantage of the
lawn. At the top of the hill, a spot Frederick Law Olmsted chose
in 1872 for its expansive views of Boston and the old trees that
stood here back then, I park my car under the shade of quaking
aspens and red maples in a lot blowing with cottonwood seeds—
duckling-soft in my palms—and march towards the Admissions
building, not to check myself in, this time, but to spend part of
my sabbatical year researching in their archives. It's taken more
than two years to get access. I cannot wait. And at the same time,
I'm anxious to have returned to a site where I was once so sick.

In New England, McLean is often thought of as a hospital for the
elite, once housing literary giants like Robert Lowell, Sylvia Plath,
Anne Sexton, musicians Ray Charles and James Taylor, and, more
recently, Selena Gomez, and NFL star Brandon Marshall. Admittedly, when it was time for me to go (my stay was entirely covered

by my free Massachusetts health insurance), I told myself this was the best place in the world for the treatment of Obsessive-Compulsive Disorder, and that its reputation and legacy meant I'd get my greatest shot at overcoming the disorder. All of that turned out to be true. I have been very lucky. My treatment team was incredibly skilled and experienced. They saved me. Or, as they say, they helped me save myself.

But today, looking back, I think that it was also this place—its beautiful landscape, the old trees, feeling apart from the world in a wind-blown sanctuary—that played a role in my healing, too, and that maybe the hospital's nineteenth-century founders deserved some credit as well. Twelve years later, when I return to campus, my first impulse is to say hello to my old friend, my favorite oak tree, which has stood rooted in the bowl since the hospital's founding. I know because all year, I've been obsessively tracing it back in time, scrutinizing old photos and lantern slides to track its growth since 1895.

McLean's trees might seem like a trivial piece of this centuries-old institution's history. But my interest is rooted (pun intended) in the hospital's most fundamental principles, which have been carried forth into the present day: the Moral Cure, a humane approach to mental healthcare that's grounded in treating patients with kindness and helping them to heal with architecture, the environment, and access to the work and hobbies that fuel their bodies, minds, and spirits. Today, in the wake of the ongoing pandemic and rising anxiety and depression, we espouse the benefits of "tree bathing" and spending time outdoors, away from our tech and devices. McLean's campus offers a living map of the history of psychiatry's value of the natural world in patient care.

I've never been good at memorizing names and classification systems, but lately, I've been trying to know natural places more intimately by learning to differentiate between species. It started five years ago with a walk in Michigan under the trees, when I asked a friend to help me. I was embarrassed that I didn't even know which were maples and oaks. A Canadian scientist, she said to think of the Canadian flag, maple syrup: a stouter leaf. Oaks, on the other hand, she said, are shaped like our hands. She stretched her fingers across a leaf. I did the same, my fingertips reaching to the edges of a leaf's lobes, mimics of each other (we both have large palms and long, thin fingers). Robin Wall Kimmerer writes about the importance of knowing species' names, even if they're only the names we make for them, distinguishing one from another.

When I was a patient at McLean, I didn't know the type of tree I'd been sitting under for weeks. I only knew it brought me comfort. The 1960s patient map I'm given over a decade later tells me it's a white oak, and when I go back now, I marvel at its leaves that reach and flutter like thousands of hands in the air.

Like the naming of species and taxonomies, diagnosis is a linguistic process based on defining characteristics. The year before I went to McLean, I'd found myself in the Alice-in-Wonderland state of being thrust into a new world because my body and past had changed with sudden knowledge. At thirty-two, I told someone my symptoms, and she gave me new words, a diagnosis: OCD, Obsessive-Compulsive Disorder. She might as well have said, Here is a different story of your life.

From the time I was eight or nine, when the intrusive thoughts began to torment me, I thought I was secretly a monster, and that one day, I'd go insane and be locked up forever. When I finally learned that I didn't have a case of secret sociopathology but a treatable anxiety disorder, I was floored—both relieved and disillusioned: I'd spent twenty-two years avoiding much of what I wanted simply because of what I now know was just OCD? How could I not have known? How could I have wasted so much time? And now that I was so sick, how on earth was I going to get better?

My life unspooled behind me in a shape I couldn't narrativize. I lost my ability to write. Isolated in my apartment, I made fabric pictures, intuitively trying to make stories out of my overwhelming experiences and feelings.

I spent a year and a half in darkness, lost, trying to do OCD treatment but feeling like it was so deeply rooted throughout my body, it was futile. Attack it in one place, it grew in another. Working hard to stay alive, I spent time outside, under the trees with my dog, and I went to look at art. Having planned my own death one winter day, I found myself at the Depressions Clinic, a day program at McLean. Finally, here I was in the sort of place I'd always believed I'd end up: the mental hospital, as I thought of it, an image that came from 1970s pictures of neglected people languishing in over-crowded halls. I had no idea that a place like McLean existed. I found not anguish but high-quality care, the halls filled with people like me, working to heal, and even, often, laughing. I survived that winter, and in May, I checked into the OCD Institute at McLean. This, I knew, was my chance. And it was terrifying: now, I'd have to face down every single fear.

On my return to McLean's archives last year, I dove into researching the history of the hospital, combing history books for accounts of its development, seeking the sources of its impact on me. In 1811, the Massachusetts General Hospital Trustees voted to open McLean Asylum for the Insane in Somerville, at the Barrell Mansion, a site selected for its bucolic attributes: fields that sloped gently towards Millers River, tall trees, and beautifully manicured gardens. Patients could be ferried between the hospitals by boat, food shipped from the farm at McLean to Mass General. According to contemporary historians, by the 1870s, McLean was forced to find a new location because it was being overrun by the expanding railroad system in Somerville. A few months later, Evelyn Battinelli, Executive Director of the Somerville Museum, corrects me.

"It wasn't just the railroads," she says. "The bigger problem was the pollution from the quickly growing meatpacking plants that were built around the asylum," she says. "The plants were disposing of carcasses in Millers River, and it got so bad, McLean had to move."

I go back to newspaper articles and accounts from the 1870s, and read that factory managers believed the tide would carry the animal remains out of the estuary. They were wrong. The river was soon overrun, in what one resident described as a "fetid" mess. Immigrants who lived in the neighborhood to work in the factories and plants were made to endure the stench and unsanitary waters until the town was mandated by the state government to fill the river and create an underground container for the carcasses—the first U.S. environmental law.

The Barrell Mansion and Cobble Hill were levelled: trees razed, the hill used to fill in the river, patients transported in 1895 under the cover of a series of carriage rides to the Belmont site. Relocation complete. "One by one," Reverend Edward G. Porter writes in the 1896 Massachusetts Historical Society Proceedings, "the natural beauties of our metropolis are giving way to the imperious demands of our commercial growth . . . this well-known eminence just over the river must not only surrender its half-dozen large and

well-built structures of brick and stone, its stately elms and its terraced gardens and orchards, but the hill itself is at once to be
levelled to make room for a network of tracks and freight yards."
Industrialization was unstoppable. Walk around the Somerville site
today, and you'll find desolate concrete flatlands dominated by the
railroad hub, high-rises moving steadily upwards all around, locals
still shaking their heads at the swiftly changing neighborhood.

The trustees asked Frederick Law Olmsted, who had just designed the mental hospital in Buffalo, to choose between three
sites. He selected Belmont for its rolling hills, old trees, "long
curves and easy glades," as he wrote, and a glorious, open horizon
over the treetops that stretched to Boston. Stand at the entrance of
the Administration building today, and you'll feel like Dr. Cowles,
the Superintendent who oversaw the campus he helped design,
looking out across the bowl, the expansive treetops (and, now, the
condominiums) towards the city.

In the course of carefully selecting this site and preparing it for
patients, the trees were thinned on Olmsted's advice, per his 1872
letter about the site, and others were planted, the ground levelled
in places and graded, working around the rocky outcroppings, to
make the place a healing environment for patients to walk and
wander. A McLean bulletin article from the 1970s notes that when
the Belmont site opened in 1895, "A variety of hardy New England
trees—cedars, oaks, maples, and hemlocks covered the 107 acres."

The buildings took three years to construct. They were designed
like a college campus, with buildings spread out across green space
between old trees, under what was called the "cottage plan," which
allowed for the separation of patients depending on severity of
illness (and level of noise), as well as space for natural interludes
outdoors for walking and relaxation. Olmsted planned for the
buildings to be connected by tunnels that rise above ground and
divide the landscape into courtyards.

Olmsted may or may not have been involved in the details of
the landscape design. On the record, that was left to Joseph H.
Curtis. Although Olmsted wrote that he dreaded ending up in an
institution, he would return to McLean as a patient for the final
five years of his life.

One of the terrible things about OCD therapy is that you have to
suffer so much to get better. It's such a cliché: the only way out

is through, but for OCD, that's how it works. Exposure-response-prevention (ERP) means you make a hierarchy of all your fears and obsessions, from least to most scary, and you slowly work your way up the list, facing down every single one. You invite the very worst things to happen. You don't allow yourself to ritualize. It's like lying down on a train track and waiting for the train to approach, forcing yourself to stay there, feeling its rumble. You cannot move. The anxiety rises and rises. Before the anxiety begins to abate, ERP feels like torture.

During breaks, I'd go back to my room intending to read a stack of books lent to me by my friend Ellen—Gretel Ehrlich, Amy Blackmarr, Janisse Ray—women who write about the natural world—then quickly fall asleep. I was always exhausted. But one day, I wandered outside. I walked across the patio, over the driveway, and into the bowl, towards a massive tree. It sits at the bottom of that easy hill. It was calling to me, graceful branches reaching out like a mother's arms: I'm here; you're safe. Close your eyes and listen to my leaves.

Leaning against the tree, I could feel its thick bark. I knew it had seen many more years than I had, that its roots had reached deep into the earth, emerging here and there in the grass like strong, gnarled fingers spanning the width of its canopy. It had a strength I longed for, a sense of certainty—something most people with severe OCD almost never experience. It was solid. And forgiving. Both at once. It was what I was striving to be.

Thinking about my time under my favorite tree, as I came to call it, reading the tree map made by patients in 1965–66, considering Olmsted's involvement, and in search of information about the materials used to construct the buildings, I knew I wanted to write an environmental history of McLean. But, my first few months in the archive, I still hadn't found much about the landscape—until one day, the archivist sent me an email saying, "I left something on your chair I think you'll be interested in." I raced to the archive in the morning to find Margie A. Lamar's "A Study of the History of Landscape Architecture" at McLean. Between the 1970s and 2022, Margie A. Lamar's paper was internally referenced in hospital publications, but never published beyond its walls.

Why did she do this research, and never publish it? Who was she, a woman researching McLean in the 1970s?

Lamar opens her paper by citing Olmsted at the end of his life, back at McLean as a patient: "they didn't carry out my plan, confound them!" She credits this quote to Laura Roper's biography of Olmsted, published in 1973 to wide acclaim for its breadth and depth.

Roper had been working on her biography of Olmsted, *F.L.O.*, since the 1950s, though it wouldn't be published until 1973. Lamar wrote her paper in 1975 as part of her work for a class on landscape architecture that she took at Harvard with Albert Fein, one of the leading experts on Frederick Law Olmsted.

On April 25, 1975, Laura Roper visited Fein's class, and told Lamar that Olmsted did design McLean's campus. Included in Lamar's paper is a reference to "a watercolor sketch, which is, quite probably, the work of Olmsted. It shows the arrangement of the patient buildings around the Administration Building spaced from 55 to 65 feet apart with a 'Pleasure Ground' area indicated, and connected walkways. It is dated November 1875 and is marked 'Study No. 2.'" Lamar defends her claim that this is likely Olmsted's work with reference to several letters he wrote to McLean's trustees and Joseph H. Curtis about the site and design.

Lamar wrote that she hoped that her paper would "contribute in some way to the appreciation and preservation of Olmsted's work at McLean Hospital." She is surprised that, as of the writing of her paper, he hasn't been recognized for his design of McLean in any publications.

After closely analyzing the differences between Olmsted's original plan and McLean's construction, completed in 1895, Lamar determines that, when he said he was dismayed at the hospital deviating from his plans, "he must have [been upset about] this difference of opinion about the placement of the buildings, their distance from each other, and their arrangement around a village-green." She says that it was Dr. Cowles who was largely responsible for "the ultimate placement of the buildings. In reading the trustees reports and the histories of McLean, it is obvious he was a very able and forceful man." In addition, she adds that mental hospitals had to be designed by "experienced medical men." This was because of the rules created by the group now known as the American Psychiatric Association, of which, she says, Cowles was president "at one time during his tenure at McLean." This may have been why Cowles was credited with the design rather than Olmsted. Lamar further suggests that Cowles and the trustees "may have felt Olmsted's original ideas reminiscent of the Buffalo State Hospital," which Olmsted had also designed and was in keeping with the state hospital aesthetic, which McLean was "trying to avoid."

In her introduction, Lamar describes the treatment of mental health in 1872 to have been shaped by "Moral Management," as defined in the late 1700s by Samuel Tuke in England and Phillipe Pinel in Paris. Patients were to be taken to "proper asylums . . . a calm retreat in the country is to be preferred. When convalescing, allow limited liberty; introduce entertaining books and conversation, exhilarating music, employment of the body in agricultural pursuits." Instead of being sent to the jails or poorhouses, Lamar writes, "[a]t last the mentally ill were to be treated with kindness in separate institutions." This was the case the Mass General trustees had so fervently made in the campaign to found McLean.

McLean offered all the benefits the Moral Cure espoused, in Somerville and then Belmont, sending patients for walks and time in glorious gardens and under the trees, and assigning them to work in the stables and on the farm, which provided food for both McLean and Mass General Hospital. In the 1960s–70s Disability Rights Movement, such work was outlawed, because it allowed institutions to take advantage of vulnerable people's free labor. But, as Oliver Sacks observed in the 1990s, this also took away many patients' sense of purpose, the worthy contributions they could make between therapy sessions and group meetings, and the chance to stay physically active in camaraderie with other patients and staff. Now, they were left to their own devices, often listless on the wards.

Lamar says she found little about the source of McLean's trees, but that Joseph Curtis, appointed one of the leads on the project by the trustees, was ". . . an engineer for several subdivisions in Belmont from 1872–1910," and might have transplanted trees slated for "removal" from those sites to McLean. Construction files at McLean reveal that in 1893, Curtis contracted with the Shady Hill Nursey Company for "deciduous trees, vines, and shrubs."

In addition to his work around Boston, Curtis was a well-known summer resident of Mt. Desert Island in Maine, where he "co-founded the Northeast Harbor Summer Colony with William Doane and Charles Eliot, then president of Harvard University," and designed several neighborhoods and gardens, including his own home, the Thuya Garden, Asticou Terrace, and a portion of

Land's End. Eliot's son became an apprentice to Olmsted and, in 1890, petitioned to save the 400-year-old Waverley Oaks, just across from McLean's campus, which Olmsted references in his 1872 letter to the trustees about the value of McLean's Waverley site. Eliot's argument resulted in the formation of The Trustees of Public Reservations, and then, in 1893, the nation's first Parks System.

Lamar writes that other sources of McLean's trees included Ebenezer Francis Bowditch, one of the trustees, who would select from his "farm in Brookline . . . choice fruits and flowers to bring to the hospital." In addition, Joseph Breck and Sons supplied some trees, shrubs, and flowers. Finally, Lamar surmises that because Olmsted's firm was working on the Arnold Arboretum as of 1880, it's possible he may have transplanted trees from there; this is the only claim without any direct evidence. She writes that "Charles Sprague Sargent, founding director of the Arnold Arboretum, . . . [was believed to have] had a role in selecting trees for McLean," but the source for this belief isn't clear.

Looking back at photos from the Waverley site, this white oak has been here since at least 1895. I hold a lantern slide up to the light of a window, squinting to see ice skaters in the bowl, next to the tree. Because of its large size even at that time, it must have been one of the trees native to this site, one of the survivors in Olmsted's plan to "thin" the wooded landscape into glades.

I take a break from hunching over the table at the archives and snapping photo after photo to document, to slip outside into the sun, and walk down the hill to the old oak. I find a broken bird's egg, the yellow yolk dried inside, and a colony of mushrooms spread out in the shade, making new life. I sit and lean against the bark, listen to the leaves, embracing what always feels, here, like respite from the rest of the world—quiet, rest. Olmsted was right. This landscape does bring patients—at least, me—a sense of tranquility.

Thirteen years ago, at a time when I was being schooled in mindfulness as a way to overcome severe anxiety, I came to this oak for respite, and for lessons. Stand here. Listen. The leaves are moving, the grass is growing, the sun and clouds are shifting above you, written in the shadows that pass over your legs. Faith is a radical act, the ability to believe in what you can't see happening. Faith not in the religious sense but in the spiritual—faith that the world is "not one long string of horrors," as the poet Gerald Stern writes, but ultimately good, that the sun will rise in the morning, that I would survive this time and emerge on the other side. My treatment providers pointed me out of the dark forest in which I was trapped, and I walked, blind, listening to their voices, following every instruction. It took all of my energy to fight, believing my team knew how to get me through. In my downtime, I sat under this tree, feeling still but knowing its cells were multiplying and expanding as I leaned on a trunk that looked and felt massive, immobile, and permanent. This tree couldn't possibly be moving. But it was.

Researchers have known for decades that walking under trees benefits our health. The pandemic and our time forced outdoors popularized the Japanese concept of shinrin-yoku or forest bathing, "walking under the trees and breathing the air" or "immersing oneself in nature, using one's senses."

A 2022 survey of the shinrin-yoku literature found that the most accurate results document the benefits for people suffering from anxiety and depression. That research is now expanding to assess the effects on people with other mental health challenges, including those at mental hospitals. Other scientists have documented the value of monoterpenes, chemicals emitted by certain plants that decrease our stress levels.

There's something satisfying about science confirming how good it feels to breathe in the air under tall trees, that shift in the body that comes after a walk in the woods. That feeling of calm is now documented: We are being changed at the molecular level.

This recent science confirms what psychiatrists and healthcare providers have known for centuries: trees benefit our health. These were all the same prescriptions given by the founders of the hospital in the days of the Moral Cure. Looking at 1880s photos from the Somerville site, we see women standing under the trees with walking sticks, waists cinched in their dresses. A group looks over the edge of a bridge at the pond below. They also spent time weaving in the loom room, looking at art in a space curated by one of the trustees, playing tennis, and ice skating. And just as women at McLean were told to do in the 1850s, when I was at McLean in 2010, we were taught to get outside, move our bodies, and practice what we call today mindfulness, focusing on the present moment. We took up hobbies we'd lost along the way.

Those traditions gained even more force in the 1960s. In the midst of the national environmental movement that had begun to take hold in the wake of publications such as Rachel Carson's *Silent Spring*, patients Ann Lord and Stewart Sanders collaborated with trustee Mrs. Charles A. Coolidge to make a remarkable map of the trees of McLean. Coolidge lived right across the street from McLean in the 1930s while raising her children with her husband, Charles Coolidge Jr., a descendant of one of the architects who designed many of McLean's buildings. Charles, like his wife, seemed to be committed to preserving the town's natural spaces, as he served on the new Belmont Conservation Commission in 1967.

Perhaps the Coolidges were inspired by the revival of interest in Olmsted's work in the 1950s and 60s, which, writes the editor of Fein's 1972 book on Olmsted, "a group of scholars . . . are chronicling and illustrating . . . for the benefit of the new environmentalists—for many of whose ideals Olmsted is the prototype." Fein writes that it's only natural that Olmsted should be revisited in the wake of the "three major crises undergone by this nation during the last decade—racial, urban, and ecological."

Today, with the ongoing Black Lives Matter movement, the mental health crisis brought on by the pandemic, and the climate crisis, we are in not so dissimilar times. We sit at another axis, the intersection of environmental and mental health research, focused on how the environmental crisis impacts all of us around the world, particularly those losing their lives and homes in the southern hemisphere. Researchers are focusing on what we can do to prevent climate change and mitigate the damage for those who are losing the most even though they hold the least responsibility for its cause.

Walking back onto McLean's campus last year was a strange experience, jolting me back to the time I stayed at NB1, sleeping on a twin mattress in a room with a teenager who was fighting similar demons. But, just as I learned that exposure therapy brings the abatement of anxiety, allowing me to finally laugh in the face of my OCD (Take that! I began to think when an exposure was done), I found that coming back to the hospital wasn't scary at all. It was full of stories. Just as McLean was designed to be a home, as Lamar writes, so it became one for me—first when I went to the OCDI, and now that I've delved into its archives and rested, again, under its trees. I have been very lucky.

For years after I left McLean, I pressed ahead, trying to make up for lost time. I had a doctorate to finish, a job to earn, tenure to race after, and finally, the exhilaration and exhaustion of single motherhood consumed me. Now, enough space has passed between the years I was sick and today, with so much other work done, that I can inhabit the space again—differently, full of curiosity.

Mama, my child asked on a recent walk through the city, when I insisted we stop and look up at a tall old oak tree, *Do you wish you were a tree?*

I laughed. It was a fair question. We spend a lot of time looking at trees together.

No, I say, *but I do love them.*

On my breaks from the archives, sometimes, I go back and visit the white oak. I sit under its ever-widening branches, marvel at the mushrooms emerging in June, peek in at the massive hollow where the trunk split apart, and then I sit and listen to its leaves for as long as I can. These days, I only have a few minutes here and there for quiet interludes.

McLean's site in Belmont is the result of the onslaught of environment damage caused by the industrial revolution's second wave, with the railroads and meatpacking plants housed at the original Somerville site. The first time I drove through the old site, now called Inner Belt Park, I saw it as desolate, a contemporary wasteland, the sad castoff of industrialization. But now, returning in early spring, I see something else: rows of trees. They're saplings, still, but someday, I hope, they're going to line these roads as elders, and someone will walk under them, and wonder how they came to arrive here. Someone will dig around for the stories. Someone will unearth their histories. Maybe. At the very least, they're offsetting carbon emission and improving our air, giving us shade and some beauty. Walk up close, and you'll see the moss and lichen already forming on their bark, as if they're all making themselves at home. This, to me, is hope.

I get up and walk away from my favorite tree, meander across campus to my car, and drive back to Somerville to wait for my child to emerge from school, into my outstretched arms.

AYANA ELIZABETH JOHNSON

Be Tenacious on Behalf of Life on Earth

FROM *Time*

CONGRATULATIONS, GRADUATES. AND congratulations to all those who love and support you and have helped you reach this milestone.

When I asked for advice on what to say today, someone replied: "Just tell the truth. Kids love that shit." Another person chimed in, "You know who else needs to hear the truth? Their parents." So this is for you too.

Here we go. The only way out is through.

In the past four years, even if you didn't do all the readings (same, don't worry), you've learned a lot about the crumbling systems of nature, of government, and of justice. An intertwined morass of crises. Now, you must use those truths to keep a fire in your belly, to fuel all the good you must do in the world.

Moments like today's celebration are not a foregone conclusion, even for the most fortunate among us. The future is not guaranteed. And, most importantly, the future is not yet written. Every single person here—you—can help shape it.

So, what I want you to hold on to today are two words: *transformation* and *possibility*.

To address the climate crisis, the all-encompassing challenge that will touch whatever life and work you will go on to, requires that we not just change or adapt, but that we *transform* society, from extractive to regenerative. This is a monumental task. And it requires that we focus not on endless analysis of the problem, but on summoning an expansive sense of *possibility*, on harnessing our imaginations and our creativity. This is not to sugarcoat the horrific scientific projections. I am a scientist; I know what we are up against. And I will spare you that litany on this, your graduation day, and just say: it's going to hit 89 degrees today, in Vermont, in May.

What this moment in history requires is a tenacious focus on *solutions*, and a vision of what we are working toward, of what is *possible*.

Much of college is about parsing literature, pondering details, and wallowing in the sweet mud of nuance. I am here to encourage you to enter this next phase of your life with as much simplicity and moral clarity as you can muster. Some things are simply right and some things are simply wrong. And the devil does not need an advocate.

It is right to steward life and justice on this magnificent planet. It is right to quickly transition to renewable energy. It is right to protect and restore habitats and species. It is right to hold corporations accountable. It is right to ensure a just transition, leaving no one behind. And it is right to enact strong government policies that will accelerate all of this.

On the other hand, it is wrong to make this magnificent planet unlivable. It is wrong for the corporations who got us into this mess to continue to profit while they set the world on fire. It is wrong to drive one *million* species extinct by changing the climate, destroying habitats, and dousing the planet with pesticides. And it is wrong to leave the most vulnerable to bear the heaviest climate impacts.

Some people might tell you that seeing stark right and wrong is naïve. As you age, it will be tempting to succumb to endless compromise as the norm. Resist. And let's be clear: moral clarity is not the same thing as naïveté. It is naïve to expect that governments and corporations will do the right thing, or that someone else will handle it. It is naïve to think we can "solve" or "stop" climate change. It is also naïve to give up, when every tenth of a degree of warming we prevent, every centimeter of sea level rise we avoid, every species we save, and every increasingly unnatural disaster we avert all matter so very much.

In 1967, Dr. Martin Luther King gave a speech against the Vietnam War, and those remarks could not be more apt five decades later:

> We are now faced with the fact that tomorrow is today. We are confronted with the fierce urgency of now. In this unfolding conundrum of life and history there is such a thing as being

too late. Procrastination is still the thief of time. . . . Over the
bleached bones and jumbled residue of numerous civilizations
are written the pathetic words: "Too late." There is an invisible book of life that faithfully records our vigilance or our
neglect. . . . Now let us begin. Now let us rededicate ourselves to
the long and bitter—but beautiful—struggle for a new world.

There is one big question that drives my work: What if we get it
right? What if we behave as if we love the future?

The truly excellent news is that we already have the solutions
we need. We don't need to wait for some jazzy new technology
to save us. We are at the moment where each of us must find our
best role in getting it right. So, what can you do? First, let's re-
frame that question: What can *we* do? How can you contribute to
an important effort? (As an aside: The absolute last option should
be to start your own nonprofit! A lack of nonprofits is not what's
holding us back. Join something, and improve it.)

Here's one way to think about what you can do. Think about a
Venn diagram with three overlapping circles.

- In the first circle is, what are you good at? What are your areas of
  expertise? What can you bring to the table? Think about your skills,
  resources, and networks. You have a lot to offer.
- The second circle is, what is the work that needs doing? Are there
  particular climate and justice solutions that you're keen on? Think
  about system-level changes and things that can replicate or scale.
  Things like composting initiatives, building retrofitting, wetland protection, adaptations for heat waves, energy conservation, and getting
  climate candidates elected. There are heaps of options.
- And the third circle is, what brings you joy? Or perhaps a better word
  is *satisfaction*. What gets you out of bed in the morning? There are so
  many things that need to be done—please do not pick something that
  makes you miserable! *This is the work of our lifetimes*, so it's imperative to
  avoid burnout. Find things that energize and enliven you.

The goal is to be in the heart of this climate action Venn diagram,
where these three circles overlap, for as much of your life as you
can. If you're familiar with the Japanese concept of Ikigai for finding your purpose, you can consider this a simplified version of that.

The quest for this sweet spot is how I ended up co-founding
Urban Ocean Lab, a policy think tank for the future of coastal

cities—as a marine biologist and Brooklyn native, who is worried about sea level rise, and finds joy in policy change. (Yes, I co-founded a nonprofit—with Jean Flemma, a Middlebury alum—but it was the last resort after years of trying to figure out another way to do that work.) It's also how I ended up co-editing the feminist climate anthology *All We Can Save*—I'm a word nerd who knows a lot of phenomenal people who are doing critical work we all need to hear about. It's how I ended up co-hosting the podcast *How to Save a Planet*, seeing a need for journalism on climate *solutions*. And how I ended up co-creating the Blue New Deal when the Green New Deal all but left out the ocean.

In all of these examples, the word "co" is key. Co-founding, co-editing, co-hosting, co-creating. Good collaboration is the exponential. Especially so because the complex challenges we face are inherently interdisciplinary.

A lesson I learned early on was: don't work with jerks. The positive inversion of that is: find your people. Someone asked me the other day, "Who are your people?" And somehow, without a moment of hesitation, directly from my soul came the words: My people are conjurers. They dream things up, make something where there was nothing, something the world needs. They don't stop at dreaming, they make magic in the real world.

It gives me goosebumps to think how lucky I am to get to say that sincerely. It may take time to find your people, and those people will probably change over the decades. But when you find them, hold on. Support them and love them. Lean into possibility together. Have a blast together doing big things that make the world better. Help reel a new world into existence.

The thing at the center of your Venn diagram can also change. It certainly has for me. And you can have multiple diagrams at once. I certainly take a portfolio approach to my work. But keep going toward where you see a need, where there are gaps you can fill. Keep conjuring.

My resume only seems like a reasonable path in hindsight—there was a lot of bushwhacking and a lot of trial and error. Trial and error is a very valid way to approach life and career! But you have to *really* try, and then really learn from the errors.

I have a feeling a lot of you will end up as trailblazers, not because you went to a great school and should be showing others the way, but because the natural world is changing so fast that a lot of the work that needs doing involves career paths that are unpaved at best.

As climate journalist Kendra Pierre-Louis puts it: Because of climate change, we are stepping into the unknown. So, yes, study the past, and learn from it, so you can avoid reinventing the wheel, but recognize that more than ever the past is an imperfect template for what we are facing.

An anecdote for those of you going on to graduate school, now or later. When I went off for my marine biology Ph.D., I told my mother it would take five or six years, and she said "Oh no!" Because the world did not have five years for me to hang out and take classes—it was important to be doing real things out in the world ASAP. I ended up designing a dissertation in collaboration with the fisheries department of the Caribbean Island of Curaçao. I did hundreds of SCUBA dives. I drove around for months with a cooler full of beer in my trunk, conducting hundreds of stakeholder interviews. (Basically, I have a Ph.D. in drinking beer with fishermen.) And I collected data the government then used to put in place new regulations to make fishing more sustainable, data they otherwise would not have had. Escape gaps are now required in fish traps, and gill nets are banned.

Another way to put all this is: be a problem solver. Regardless of the field you choose, whether it's climate finance, engineering, urban planning, environmental law, teaching, or transportation. There is nothing sexier than a problem solver. And be an influencer—not in the sexy Instagram model way, but in leading by example. There is a soft power in how you spend your time and money, how you look out for each other, how you contribute to making solutions happen.

I did not use AI to help me write these remarks. But I did ask social media what I should say to you, and got some excellent replies. Let this be a lesson in the wisdom of the *human* collective.

- The people in power do not always do the important work. Often, important work is left to regular people like us.
- There is no "formula" for what your life or career path looks like. Failures are your fertilizer.
- Be kind, to yourself and others. Give a shit. Always vote.
- Every job can be a climate job! No matter if you studied econ or were premed.

- You do *not* need an opinion on every topic. If you don't know enough to form one, have humility. And ask questions.
- Go outside, and don't come back in until you've fallen in love with the Earth.
- From Middlebury distinguished scholar Bill McKibben, who co-founded climate organization 350.org with Middlebury students: It's actually more fun to try to change things than just to go along.
- And last: Don't graduate, commence.

Excellent advice. And there were a hundred more. People are so excited to welcome you into the "real world."

There's a piece of advice I'm not going to give you. I'm not going to tell you to follow your heart. In part because I have no idea what's in your heart, so that feels a bit reckless, to be honest. But also because we live in a pivotal time for preserving life on this magnificent planet. So I can't in good conscience tell you to "follow your passions and then everything will magically fall into place"—I mean, who comes up with that stuff?! Instead, I suggest: go where there is need and where your heart can find a home.

Honestly, it feels much conventional wisdom doesn't apply. Like, it's okay to sweat the small stuff. Details matter. Especially early in my career, much of my success was due to being assiduous about the quality of anything I put my name on. And, FYI, you may have gotten good grades in college, but no one will care about your GPA after today. You will have to *earn* your professional reputation from scratch.

For the last four years, you've had a gorgeous liberal arts education, you've been training your mind and your social abilities. Maybe it's been individualistic, and today is about celebrating *your* achievements. But even today, you are part of a whole. So when the dust settles on this big occasion, thank the people who got you here today. Actually go write thank you notes, on paper, or pick up the phone. It will be cherished. Sometimes a text simply will not do.

And then, for the rest of your lives, you must find ways to be part of something meaningful, something beyond yourself. The grand goal, as I see it, is to be useful. Making useful contributions to the world, to solving important problems, is so deeply gratifying—and punctuated with joy and delight!

You have sooooo much more to learn. It's going to be fascinating. So, respect your elders, they have a lot to teach you. But also, grownups clearly do not have it all figured out. Not by a long shot. So we need your help.

Possibility

Transformation

What you do matters

How you show up matters

That you show up matters

How you live matters

How you love matters

What my parents said to me repeatedly, as I was initially charting my path, I will now say to you: To whom much is given, much is expected. You must give back.

Be tenacious on behalf of life on Earth.

You were made for this moment.

Contributors' Notes

ANNIE GOWEN is a correspondent for the *Washington Post*'s National desk. In 2023, she co-authored the *Post*'s series "The Human Limit," which examined the health consequences of climate change around the globe. She was the India bureau chief from 2013 to 2018. Her long-form journalism was honored as part of the *Best American Newspaper Narratives*, and she was part of a team whose coverage of mass shootings won a Scripps Howard Award and was a finalist for the Pulitzer Prize for Breaking News Reporting in 2020. As the *Post*'s India bureau chief, she covered India, other South Asian nations, and Myanmar, and twice won the Daniel Pearl Award for reporting on South Asia.

RAYMOND ZHONG covers climate and the environment for the *New York Times*.

SAMMY ROTH is the climate columnist for the *Los Angeles Times*. He writes the twice-weekly Boiling Point newsletter and focuses on clean energy solutions. He previously reported for the *Desert Sun* and *USA Today*.

SARAH KAPLAN is a climate reporter at the *Washington Post* covering humanity's response to a warming world.

DOUGLAS FOX (www.douglasfox.org) is a freelance journalist who writes about earth, biology, and polar sciences. Doug has written for *Scientific American*, *National Geographic*, *The Atlantic*, *Virginia Quarterly Review*, *bioGraphic*, *Science News*, and other publications. He is a contributing author to *The Science Writers' Handbook* (Da Capo, 2013).

ALEX CUADROS is the author, most recently, of *When We Sold God's Eye: Diamonds, Murder, and a Clash of Worlds in the Amazon.*

AMANDA GEFTER is a freelance science writer specializing in fundamental physics, cognitive science, and the philosophy and history of science. Her writing has appeared in *Nautilus, The New Yorker,* the *New York Times, Quanta, Nature, Scientific American, The Atlantic,* and *New Scientist,* among others. She was a Knight Science Journalism Fellow at MIT and is the author of *Trespassing on Einstein's Lawn.*

NICK BOWLIN is a freelance writer and a contributing editor at *High Country News.* His work has appeared in *ProPublica, Harper's Magazine, The Drift,* and *The Nation,* among others. He lives in Gunnison, Colorado.

HEIDI BLAKE joined *The New Yorker* as a contributing writer in 2022, after working as an investigative reporter and editor at BuzzFeed News and the London *Sunday Times.* She is the author of *From Russia with Blood: The Kremlin's Ruthless Assassination Program and Vladimir Putin's Secret War on the West,* building on an investigation at BuzzFeed News that was a Pulitzer finalist and won the IRE's Tom Renner Award in 2018. She is also the co-author of *The Ugly Game: The Corruption of FIFA and the Qatari Plot to Buy the World Cup,* drawing on the revelations of hundreds of millions of documents leaked from inside FIFA.

CAROLYN KORMANN writes about science, the environment, food, and climate change for *The New Yorker* magazine, among other publications. Her work has been recognized with a best-food-coverage award from the James Beard Foundation, and has been previously noted by the Best American series, in its *Science and Nature Writing* and *Travel Writing* collections. She is a 2024 writer-in-residence at the Virginia Center for the Creative Arts, a 2023 MacDowell Fellow, and a 2019 recipient of an Abe Fellowship from the Social Science Research Council. Her first book, *How to Be a Bat,* is forthcoming. She lives in Marfa, Texas, and New York City.

IAN FRAZIER writes essays and long nonfiction. He lives in Montclair, New Jersey.

BEN GOLDFARB is an environmental journalist whose work has appeared in *National Geographic, The Atlantic, Smithsonian, Science,* and many other publications. He is the author of *Crossings: How Road Ecology Is Shaping the Future of Our Planet,* named one of the best books of 2023 by the *New York Times,* and *Eager: The Surprising, Secret Life of Beavers and Why They Matter,* winner of the PEN/E. O. Wilson Literary Science Writing Award. He lives in Colorado with his wife, Elise, and his dog, Kit—which is, of course, what you call a baby beaver.

BRENDAN EGAN's poetry and fiction have appeared in *Threepenny Review, Catapult, Rattle, Witness,* and other publications. A graduate of the Department of Dramatic Writing at New York University and the MFA program at McNeese State University, he lives in West Texas with his wife, Stacy, and their two young children. He teaches at Midland College and keeps a pollinator garden.

B. M. OWENS is a poet and essayist who recently moved to Bellingham, Washington. Owens grew up in South Florida and received her MFA in poetry from Florida International University in 2022. Her work has been nominated for *Best New Poets* and the Pushcart Prize and can be found in *Salamander, Cherry Tree,* and *Silk Road Review,* among other journals. Her debut chapbook, *Don't Be Another Girl,* was a semi-finalist in the New Women's Voices Chapbook Competition and published by Finishing Line Press in 2024.

DAN MUSGRAVE was raised by animals in rural Kansas. Before turning to writing and photography, he spent nearly seven years doing linguistic, cognitive, and behavioral research with bonobos while they taught him how to be a better person.

ELIZABETH KOLBERT is a staff writer at *The New Yorker.* She's the author of several books, including *The Sixth Extinction,* which won a Pulitzer Prize, and *Under a White Sky,* named one of the ten best books of the year by the *Washington Post.* She is the recipient of a Heinz Award and recently received the BBVA Biophilia Award for Environmental Communication.

ISOBEL WHITCOMB is a Portland-based science and environmental journalist whose work explores the relationship between humans and the natural world. Their writing appears in *Atmos, Sierra, Slate,* and more.

JOE SPRING is an online science editor at *Smithsonian* magazine and a former editor at *Outside* magazine. He has written for *Smithsonian, Outside,* the *New York Times, Hakai,* and others.

EMMA MARRIS writes about human-nonhuman entanglement from Portland, Oregon. Her book on changing relationships between humans and animals, *Wild Souls,* came out in July 2021.

LINDSAY RYAN is an internal medicine and HIV doctor in San Francisco. Her essays and articles have appeared in the *Washington Post, The Atlantic,*

Slate, and *JAMA.* She holds an MD from Harvard Medical School and an MFA in nonfiction from the Bennington Writing Seminars.

RACHEL MAY is the author of four books of fiction, nonfiction, and image+text. Her essays have appeared in the *New York Times, Guernica, Boston Globe Magazine, Outside, National Geographic, Sierra, Wellcome Collection,* and others, and she's been a resident and fellow at the Millay Colony, Vermont Studio Center, and VCCA.

DR. AYANA ELIZABETH JOHNSON is a marine biologist, policy expert, and writer. She is co-founder of the nonprofit think tank Urban Ocean Lab, co-editor of the bestselling climate anthology *All We Can Save,* and author of *What If We Get It Right?: Visions of Climate Futures.* She is in love with climate solutions.

Other Notable Science and Nature Writing of 2023

Izzy Acevedo
  Sakura Storytellers. *Arnoldia*, April 1, 2023.
Ross Andersen
  The Aftermath of a Mass Slaughter at the Zoo. *The Atlantic*, March 2, 2023.
Philip Ball
  How Life Really Works. *Nautilus*, November 6, 2023.
Matthew Battles
  What the Meadow Remembers. *Arnoldia*, September 6, 2023.
Meg Bernhard
  The Enigmatic Method. *VQR*, June 11, 2023.
Willy Blackmore
  Social Dilemma: What's at Stake When We Propel Wild Birds to Stardom? *Audubon Magazine*, October 3, 2023.
Paul Bogard
  We're Watching the Sky as We Know It Disappear. *New York Times*, June 20, 2023.

Eric Boodman
  In a Town Plagued by an Environmental Crisis, a Local Abortion Debate Consumes Public Attention. *STAT*, April 18, 2023.
Alice Callahan
  Her Doctor Said Her Illness Was All in Her Head. This Scientist Was Determined to Find the Truth. *New York Times*, March 14, 2023.
Craig Childs
  Glen Canyon Revealed. *High Country News*, February 1, 2023.
Tressie McMillan Cottom
  Ozempic Can't Fix What Our Culture Has Broken. *New York Times*, October 9, 2023.
Christopher Cox
  The Trillion-Gallon Question. *New York Times Magazine*, June 22, 2023.
Jess Craig
  Gorillas in Their Midst. *bioGraphic*, November 22, 2023.

Marion Renault
Hiccups Have a Curious
Connection to Cancer. *The
Atlantic*, April 3, 2023.
Marion Renault
Animals Are Dying in Droves.
What Are They Telling Us? *The
New Republic*, May 3, 2023.
Tim Requarth
Pour One Out. *Slate*, April 23,
2023.
Fletcher Reveley
Scientists Warned of a Salton
Sea Disaster. No One Listened.
Undark, July 3, 2023.
Laura Sanders
Lifting Depression with
Brain Implants. *Science News*,
September 23, 2023.
Michael Schulson
For a Pivotal Vaccine: Trial,
Error, and Two Young Lives.
Undark, October 9, 2023.
Sonia Shah
The Case for Free-Range Lab
Mice. *The New Yorker*, February
18, 2023.
Lauren Silverman
Zero Lead Is an Impossible
Ask for American Parents. *The
Atlantic*, August 23, 2023.
Richard Sima
A Catatonic Woman Awakened
After 20 Years. Her Story May
Change Psychiatry. *Washington
Post*, June 1, 2023.
Manvir Singh
Red Shift. *The New Yorker*,
September 25, 2023.

Laura J. Snyder
A Kingdom of Little Animals. *The
American Scholar*, June 1, 2023.
James Somers
Begin End. *The New Yorker*,
November 13, 2023.
Julia Sonenshein
The Climate Is Changing
Everything—Including Our Sex
Lives. *The New Republic*, July 27,
2023.
Ashley Stimpson
The Race to Preserve the
250-Year-Old Monuments that
Mark the Mason-Dixon Line.
Popular Mechanics, October 19,
2023.
Jia Tolentino
The Ozempic Era. *The New
Yorker*, March 20, 2023.
Leath Tonino
The Lookout. *Adventure Journal
Quarterly*, September 1, 2023.
Leath Tonino
Did You Smell the Dirt Before
You Left Home? *Outside*, May 1,
2023.
Boyce Upholt
Monuments upon the
Tumultuous Earth. *Emergence*,
April 23, 2023.
Boyce Upholt
The Unending Quest to Build
a Better Chicken. *Noema*,
December 20, 2023.
Moises Velasquez-Manoff
The Mystery of My Burning
Esophagus. *New York Times
Magazine*, October 4, 2023.

Nicole Walker
Everyone Wants to Be a
Brine Shrimp. Terrain.org,
December 29, 2023.
Emily Witt
Fertile Ground. *The New
Yorker*, April 17, 2023.
Katherine J. Wu
Go Ahead, Try to Explain
Milk. *The Atlantic*, June 22,
2023.
Kate Yoder
The Environmental Disaster
Lurking Beneath Your
Neighborhood Gas Station.
Grist, June 14, 2023.

Ed Yong
Fatigue Can Shatter a Person.
The Atlantic, July 27, 2023.
Andrew Zaleski
What Does Wellness Mean When
You're Living with an Incurable
Disease? *GQ*, February 6, 2023.
Andrew Zaleski
On the Hunt for America's
Forgotten Apples. *Outside*,
May 18, 2023.

ABOUT

MARINER BOOKS

Mariner Books traces its beginnings to 1832 when William Ticknor cofounded the Old Corner Bookstore in Boston, from which he would run the legendary firm Ticknor and Fields, publisher of Ralph Waldo Emerson, Harriet Beecher Stowe, Nathaniel Hawthorne, and Henry David Thoreau. Following Ticknor's death, Henry Oscar Houghton acquired Ticknor and Fields and, in 1880, formed Houghton Mifflin, which later merged with venerable Harcourt Publishing to form Houghton Mifflin Harcourt. HarperCollins purchased HMH's trade publishing business in 2021 and reestablished their storied lists and editorial team under the name Mariner Books.

Uniting the legacies of Houghton Mifflin, Harcourt Brace, and Ticknor and Fields, Mariner Books continues one of the great traditions in American bookselling. Our imprints have introduced an incomparable roster of enduring classics, including Hawthorne's *The Scarlet Letter*, Thoreau's *Walden*, Willa Cather's *O Pioneers!*, Virginia Woolf's *To the Lighthouse*, W.E.B. Du Bois's *Black Reconstruction*, J.R.R. Tolkien's *The Lord of the Rings*, Carson McCullers's *The Heart Is a Lonely Hunter*, Ann Petry's *The Narrows*, George Orwell's *Animal Farm* and *Nineteen Eighty-Four*, Rachel Carson's *Silent Spring*, Margaret Walker's *Jubilee*, Italo Calvino's *Invisible Cities*, Alice Walker's *The Color Purple*, Margaret Atwood's *The Handmaid's Tale*, Tim O'Brien's *The Things They Carried*, Philip Roth's *The Plot Against America*, Jhumpa Lahiri's *Interpreter of Maladies*, and many others. Today Mariner Books remains proudly committed to the craft of fine publishing established nearly two centuries ago at the Old Corner Bookstore.

EXPLORE THE REST OF THE SERIES!